Jos A.E. Spaan, Ruben Coronel,
Jacques M.T. de Bakker and
Antonio Zaza (Eds.)

Series Editor: Joachim H. Nagel

Series in Biomedical Engineering

Editor-in-Chief

Prof. Dr. Joachim H. Nagel
Institute of Biomedical Engineering
University of Stuttgart
Seidenstrasse 36
70174 Stuttgart
Germany
E-mail: jn@bmt.uni-stuttgart.de

The International Federation for Medical and Biological Engineering, IFMBE, is a federation of national and transnational organizations representing internationally the interests of medical and biological engineering and sciences. The IFMBE is a non-profit organization fostering the creation, dissemination and application of medical and biological engineering knowledge and the management of technology for improved health and quality of life. Its activities include participation in the formulation of public policy and the dissemination of information through publications and forums. Within the field of medical, clinical, and biological engineering, IFMBE's aims are to encourage research and the application of knowledge, and to disseminate information and promote collaboration. The objectives of the IFMBE are scientific, technological, literary, and educational.

The IFMBE is a WHO accredited NGO covering the full range of biomedical and clinical engineering, healthcare, healthcare technology and management. It is representing through its 58 member societies some 120.000 professionals involved in the various issues of improved health and healthcare delivery.

Previous Editions:
Spaan, J. (Eds.): BIOMED, Biopacemaking, 2007, **ISBN** 978-3-540-72109-3

Jos A.E. Spaan, Ruben Coronel,
Jacques M.T. de Bakker and Antonio Zaza (Eds.)

Biopacemaking

Springer

Editors

Prof. J.A.E Spaan
Academic Medical Center
Department of Medical Physics and Information
Meibergdreef 9
1105 AZ Amsterdam
The Netherlands
E-mail: MBEC@amc.nl

Dr. Ruben Coronel
Department of Experimental Cardiology
Academic Medical Center
Meibergdreef 9
1105 AZ Amsterdam
The Netherlands
E-mail: coronel@amc.nl

Guest Editors

Dr. Jacques M. T. de Bakker
Academic Medical Center
Department of Experimental Cardiology
Center for Heart Failure Research
Meibergdreef 9
1105 AZ Amsterdam
&
The Heart Lung Center
University Medical Center
Utrecht
&
The Interuniversity Cardiology
Institute of the Netherlands
Utrecht
The Netherlands
E-mail: j.m.debakker@amc.uva.nl

Dr. Antonio Zaza
Dipartimento di Biotecnologie e Bioscienze
Università di Milano-Bicocca
Piazza della Scienza 2
20126 Milano
Italy
E-mail: antonio.zaza@unimib.it

Originally published as Volume 45, Number 2 in the journal Medical and Biological Engineering and Computing, 2007.

ISSN print edition: 1615-3871
ISSN electronic edition: 1860-0794

ISBN 978-3-642-09127-8 e-ISBN 978-3-540-72110-9

The Series in Biomedical Engineering is an official publication of the International Federation for Medical and Biological Engineering.

Springer is a part of Springer Science+Business Media

springer.com

Softcover reprint of the hardcover 1st edition 2007

IFMBE

The International Federation for Medical and Biological Engineering (IFMBE) was established in 1959 to provide medical and biological engineering with a vehicle for international collaboration in research and practice of the profession. The Federation has a long history of encouraging and promoting international cooperation and collaboration in the use of science and engineering for improving health and quality of life.

The IFMBE is an organization with membership of national and transnational societies and an International Academy. At present there are 53 national members and 5 transnational members representing a total membership in excess of 120 000 professionals worldwide. An observer category is provided to groups or organizations considering formal affiliation. Personal membership is possible for individuals living in countries without a member society. The IFMBE International Academy includes individuals who have been recognized for their outstanding contributions to biomedical engineering.

Objectives

The objectives of the International Federation for Medical and Biological Engineering are scientific, technological, literary, and educational. Within the field of medical, clinical and biological engineering its aims are to encourage research and the application of knowledge, to disseminate information and promote collaboration.

In pursuit of these aims the Federation engages in the following activities: sponsorship of national and international meetings, publication of official journals, cooperation with other societies and organizations, appointment of commissions on special problems, awarding of prizes and distinctions, establishment of professional standards and ethics within the field as well as other activities which in the opinion of the General Assembly or the Administrative Council would further the cause of medical, clinical or biological engineering. It promotes the formation of regional, national, international or specialized societies, groups or boards, the coordination of bibliographic or informational services and the improvement of standards in terminology, equipment, methods and safety practices, and the delivery of health care.

The Federation works to promote improved communication and understanding in the world community of engineering, medicine and biology.

Activities

Publications of the IFMBE include: the journal Medical and Biological Engineering and Computing, the electronic magazine IFMBE News, and the Book Series on Biomedical Engineering. In cooperation with its international and regional conferences, IFMBE also publishes the IFMBE Proceedings Series. All publications of the IFMBE are published by Springer Verlag.

Every three years the IFMBE hosts a World Congress on Medical Physics and Biomedical Engineering in cooperation with the IOMP and the IUPESM. In addition, annual, milestone and regional conferences are organized in different regions of the world, such as Asia Pacific, Europe, the Nordic-Baltic and Mediterranean regions, Africa and Latin America.

The administrative council of the IFMBE meets once a year and is the steering body for the IFMBE. The council is subject to the rulings of the General Assembly, which meets every three years.

Information on the activities of the IFMBE are found on its web site at: http://www.ifmbe.org.

Contents

Biopacemaking: Clinically Attractive, Scientifically a Challenge

Jacques M.T. de Bakker[1,2,3] and Antonio Zaza[4]

[1] Department of Experimental Cardiology, Heart Failure Research Center, Academic Medical Center, Meibergdreef 9, 1105AZ Amsterdam, The Netherlands
j.m.debakker@amc.uva.nl
[2] The Heart Lung Center, University Medical Center, Utrecht, The Netherlands
[3] The Interuniversity Cardiology Institute of the Netherlands, Utrecht, The Netherlands
[4] Dipartimento di Biotecnologie e Bioscienze, Università di Milano Bicocca, Milano, Italy
antonio.zaza@unimib.it

This special issue gives an overview of the current state-of-the-art of creating a bio-engineered pacemaker. The subject has potential clinical interest. Indeed, electronic pacemakers currently available have several limitations, among which inadequate rate adaptation to physiological needs, problems related to the stimulating and sensing leads and infection of the pacemaker pocket, which might be overcome by a bio-pacemaker. Generation of a bio-pacemaker has also scientific interest, because it may answer the longstanding question of whether the complex structure of the sinus node is indeed a prerequisite for reliable pacemaking, or simpler structures might work as well.

Knowledge of normal pacemaker physiology provides the ground for the development of bio-pacemakers. Various ionic currents contribute to sinoatrial (SA) node pacemaking; moreover, the sinus node comprises morphologically and functionally distinct cell types, with different intrinsic rates and response to autonomic agonists. As outlined by Opthof [11], these differences are relevant to the width and stability of autonomic modulation of sinus rate. The question may be asked of whether such complexity, probably the result of evolutionary adaptations, would also be required to create a bio-pacemaker. For reasons of practicality, the strategies proposed thus far have adopted a conservative ''one channel'' approach; however, as genetic manipulation techniques improve, reports of bio-pacemakers based on combinations of mechanisms start appearing in the literature [3].

Current approaches to bio-pacemaker generation have developed along two main lines. The first aims to induce pacemaker activity in normally quiescent (''working'') myocardium. The second involves myocardial implant of exogenous cells, engineered to sustain pacemaker activity (''cell-based'' approach) once electrically connected to the host myocardium.

Pacemaking can be induced in working myocardial cells by modification of their pattern of expression of membrane currents. The required genetic modification is usually carried out by gene transfer to the site of interest. This can be achieved theoretically by direct transfection of a plasmid incorporating the gene, or by infecting

J.A.E. Spaan et al. (Eds.): Biopacemaking, BIOMED, pp. 1–5, 2007.
springerlink.com

the tissue with a viral vector containing it. In practical terms, only the viral infection provides adequate transduction efficiency and is universally adopted. Nonetheless, the infection procedure involves a number of technical and safety problems. The replication-deficient adenovirus is a safe and practical vector; however, because the gene is not incorporated into the genome, its expression is only transient. Retroviruses, like the widely used lentivirus, incorporate the added gene into the genome, which results in stable gene expression. However, genomic transduction carries potential carcinogenic risk, which might make this type of vector less suitable for therapeutic use. Such problems have prompted the development of cell based approaches, in which pacemaker function is intrinsic to the implanted cell, or can be obtained by genetic modification prior to implant in the host myocardium.

In the cell-based approach, several strategies have been proposed. In one case spontaneously beating clusters of myocytes derived from human embryonic stem cells (hESCs) were directly used as pacemaker elements [15]. However, once implanted, these cells could further differentiate into quiescent elements, thus compromising pacemaker stability. Another, more promising, approach is based on in vitro genetic modification of exogenous cells, originally devoid of pacemaker activity, which are stably transduced with a gene encoding the current of interest. Once implanted, the modified cells electrically couple to the surrounding myocardium, and modulate its electrical activity [2, 12]. Cell-to-cell coupling is mediated by connexins, protein channels that allow ionic current flow between adjacent cells. Connexins are at hand in many cell types, including stem cells, which can successfully couple to cardiac myocytes. Success of the cell-based approach depends on the possibility of avoiding immunological rejection of the implant; thus an autologous origin of the implanted cells is highly desirable. Stem cells may be particularly suitable for generating a bio-pacemaker because they can be autologous and they replicate, thus allowing amplification of the cell population. An alternative may be the development of replicating cell-lines, engineered to achieve immunocompatibility.

To create a bio-pacemaker, the following strategies are currently followed: (1) suppression of repolarizing currents to unmask latent pacemaker currents in normally quiescent myocardial cells; (2) over-expression of a pacemaker (depolarizing) current in electrically quiescent cells to convert them into pace-making elements; (3) modulation of the expression of receptors involved in the regulation of pacemaker currents [5].

The first approach, a pioneering one in the field of bio-pacemakers, relies on the idea that ventricular ''working'' myocardium has latent pacemaker activity, but spontaneous depolarization is normally suppressed by a large repolarizing conductance, available at diastolic potential. Such a conductance is provided by the ''inward rectifier'' potassium current I_{K1}, known for its strong expression in electrically quiescent cells of the atrial and ventricular working myocardium, but virtually absent from the AV and SA node. Therefore, suppression of I_{K1} is a putative approach for creating a bio-pacemaker. The group of Marban [9] provided a proof of this concept by using a dominant-negative Kir2.1 construct, packaged into an adenoviral vector. Once infected with the vector in vivo, ventricular myocardium showed 80% I_{K1} suppression and developed automatic activity [9]. Although conceptually innovative, such an approach is encumbered by the problems related to all viral transduction methods; moreover, strict delimitation of the infection site is difficult and diffusion of I_{K1} suppression throughout the ventricle may entail pro-arrhythmic risk.

With regard to over-expression of depolarizing currents, much attention has been given to the main depolarizing current that induces spontaneous activity in the SA node, the funny current I_f, which is mediated by a family of hyperpolarization activated cyclic nucleotidegated (HCN) channels. This area has been pioneered by Rosen et al. [13], whose review in this issue summarizes the evolution of the concept and the results obtained. These authors used HCN2 as the I_f encoding isoform because the resulting current kinetics are more favorable than with HCN4 and its cAMP responsiveness is greater than that of HCN1. These investigators initially showed suitability of I_f over-expression by injecting HCN2 encoding adenoviral vectors into the left atrium or the left bundle branch of intact dog hearts. Both injection sites proved to be successful in generating an ectopic rhythm. In addition, the experiments also proved that pacemaker activity generated by expression of HCN2 was autonomically regulated. To overcome the problems related to the adenoviral infection method, the same group developed a cell-based approach. Human mesenchymal stem cells (hMSC), loaded with the HCN2 gene, were injected epicardially into the left ventricular free wall and resulted into an idioventricular rhythm at the injection site. This rhythm was significantly faster than the escape rhythm following AV nodal ablation, thus providing efficient pacemaker activity. Recent studies also explored the feasibility to convert quiescent ventricular myocytes into pacemakers using somatic cell fusion [4, 8]. Chemically induced fusion between myocytes and syngeneic fibroblasts that had been engineered to express pacemaker ion channels, has been attempted. The advantage of this approach, with respect to classical cell-based therapy, is that the gapjunctional coupling between donor cells and host myocardium, which might be suboptimal or unstable in time, is avoided.

Interestingly, a cell-based approach has also been proposed as a mean to down regulate heart rate. De Boer et al. [2] reduced beating rate of spontaneously active neonatal rat cardiomyocytes by coculturing them with I_{K1} overexpressing human embryonic kidney cells (HEK, transduced with Kir2.1 gene). These investigators also showed that the influence of Kir2.1 expressing cells on beating rate could be lessened by the application of $BaCl_2$, that blocks I_{K1}. Since pacemaker down-regulation occurred through electronic interaction between the two cell types, this result also implies that efficient connexin-mediated cell-to-cell coupling spontaneously develops between HEK cells and ventricular myocytes.

Recent evolutions in bio-pacemaking techniques involve the expression of ‘‘synthetic’’ pacemaker channels, obtained by modification of genes originally encoding non-pacemaker currents. The rationale of this approach is the concern that coassembly of added HCN proteins with those naturally expressed by the cell may result in unpredictable channel properties. To generate a synthetic pacemaker channel, Kashiwakura et al. [7] converted the depolarization activated potassium channel Kv1.4 into a hyperpolarization-activated non-selective channel by 4 point-mutations. The properties of the synthetic channel were similar to those of HCN ones, but co-assembly between endogenous and added proteins was avoided.

A requirement for successful propagation of pacemaker activity is an appropriate match between the pacemaker generator properties and the electrical load imposed by the tissue to be excited. Electrical-coupling is required for propagation between pacemaker and follower cells but, if the load is excessive, it may arrest the pacemaker by clamping pacemaker cells to hyperpolarized resting membrane potentials. The SA node has special means to circumvent this problem, including expression of an hyperpolarization-activated depolarizing current (I_f) [10] and a complex architecture of the node atrium interface [1, 6]. In the case of bio-pacemakers, the interface architecture can be hardly controlled; thus, for their development, prediction of the interplay between polarizing and depolarizing currents and quantitative estimates of the required generator size may be necessary. As reviewed by Wilders in this issue [14], accurate computer models of the SA node activity, now available, may help in understanding how depolarizing and repolarizing currents interact and respond to perturbing conditions. The problem of the match between generator and load is illustrated in this issue by Joyner et al. [6]. These investigators addressed this problem with a mixed approach in which SA electrical activity, generated by a numerical model, was electrically coupled through a variable resistor to a real atrial myocyte. This allowed to test how coupling resistance may affect the pacemaker load interaction and to obtain a quantitative evaluation of the conditions required for propagated pacemaking [6].

As highlighted in this issue, research in the field of bio-pacemaking is blooming. Nonetheless, in light of the performance and safety of the electronic pacemakers now available, development of a better alternative is an extremely demanding task. It yet has to be proven that the bio-pacemaker surpasses its electronic counterpart with regard to adaptability to physiological requirements of the body and longevity. While potentially effective pacemaking strategies have been identified, the development of genetic engineering methods suitable to implement them with safety and stability remain a considerable challenge. The possibility of uncontrolled gene expression, carcinogenic risk of viral vectors affording stable transduction and immune rejection of implants are among the problems that need to be solved before bio-pacemaking can be considered for clinical use. Moreover, ventricular resynchronization, a major advancement of artificial pacemaking, may be difficult to achieve with bio-pacemakers.

Despite these concerns, bio-pacemaking seems more easily achievable than other potential applications of cardiac cell therapy. This is because bio-pacemaking aims to restore a single function with a well-defined mechanism, it requires myocardial homing of a limited number of cells and a localized intervention. Development of bio-pacemakers may be an ideal challenge for the approach typical of bioengineering, based on a close interaction between expertise in biophysics, molecular and cell biology.

References

1. Anghel TM, Pogwizd SM (2006) Creating a cardiac pacemaker by gene therapy. Med Biol Eng Comput 45:145–155
2. de Boer TP, van Veen TA, Houtman MJ, Jansen JA, van Amersfoorth SC, Doevendans PA, Vos MA, van der Heyden MA (2006) Inhibition of cardiomyocyte automaticity by electrotonic application of inward rectifier current from Kir2.1 expressing cells. Med Biol Eng Comput 44:537–542

3. Cho HC, Kashiwakura Y, Marban E (2005) Conversion of non-excitable cells to self-contained biological pacemakers. Circulation 112(17):II-307
4. Cho HC, Kashiwakura Y, Marban E (2005) Creation of a biological pacemaker by cell fusion. Circulation 112(17): II-307
5. Edelberg JM, Aird WC, Rosenberg RD (1998) Enhancement of murine cardiac chronotropy by the molecular transfer of the human beta2 adrenergic receptor cDNA. J Clin Invest 101:337–343
6. Joyner RW, Wilders R, Wagner MB (2006) Propagation of pacemaker activity. Med Biol Eng Comput 45:177–187
7. Kashiwakura Y, Cho HC, Barth AS, Azene E, Marban E (2006) Gene transfer of a synthetic pacemaker channel into the heart: a novel strategy for biological pacing. Circulation 114:1682–1686
8. Marban E, Cho HC (2006) Creation of a biological pacemaker by gene or cell-based approaches. Med Biol Eng Comput 45:133–144
9. Miake J, Marban E, Nuss HB (2003) Functional role of inward rectifier current in heart probed by Kir2.1 over expression and dominant-negative suppression. J Clin Invest 111:1529–1536
10. Noble D, Denyer JC, Brown HF, DiFrancesco D (1992) Reciprocal role of the inward currents Ib, Na and If in controlling and stabilizing pacemaker frequency of rabbit sinoatrial node cells. Proc R Soc Lond B 250:199–207
11. Opthof T (2006) Embryological development of pacemaker hierarchy and membrane currents related to the function of the adult sinus node. Implications for autonomic modulation of biopacemakers. Med Biol Eng Comput 45:119–132
12. Potapova I, Plotnikov A, Lu Z, Danilo P Jr, Valiunas V, Qu J, Doronin S, Zuckerman J, Shlapakova IN, Gao J, Pan Z, Herron AJ, Robinson RB, Brink PR, Rosen MR, Cohen IS (2004) Human mesenchymal stem cells as a gene delivery system to create cardiac pacemakers. Circ Res 94:952–959
13. Rosen MR, Brink PR, Cohen IS, Robinson RB (2006) Biological pacemakers based on I_f. Med Biol Eng Comput 45:157–166
14. Wilders R (2006) Computer modelling of the sinoatrial node. Med Biol Eng Comput 45:189–207
15. Xue T, Cho HC, Akar FG, Tsang SY, Jones SP, Marban E, Tomaselli GF, Li RA (2005) Functional integration of electrically active cardiac derivatives from genetically engineered human embryonic stem cells with quiescent recipient ventricular cardiomyocytes: insights into the development of cell-based pacemakers. Circulation 111:11–20

Embryological Development of Pacemaker Hierarchy and Membrane Currents Related to the Function of the Adult Sinus Node: Implications for Autonomic Modulation of Biopacemakers

Tobias Opthof

Experimental and Molecular Cardiology Group, Academic Medical Center,
Meibergdreef 9, 1105 AZ Amsterdam, The Netherlands
t.opthof@med.uu.nl
Department of Medical Physiology, University Medical Center Utrecht,
Utrecht, The Netherlands

Abstract. The sinus node is an inhomogeneous structure. In the embryonic heart all myocytes have sinus node type pacemaker channels (I_f) in their sarcolemma. Shortly before birth, these channels disappear from the ventricular myocytes. The response of the adult sinus node to changes in the interstitium, in particular to (neuro)transmitters, results from the interplay between the responses of all of its constituent cells. The response of the whole sinus node cannot be simply deduced from these cellular responses, because all cells have different responses to specific agonists. A biological pacemaker will be more homogeneous. Therefore it can be anticipated that tuning of cycle length may be problematic. It is discussed that efforts to create a biological pacemaker responsive to vagal stimulation, may be counterproductive, because it may have the potential risk of 'standstill' of the biological pacemaker. A normal sinus node remains spontaneously active at high concentrations of acetylcholine, because it has areas that are unresponsive to acetylcholine. The same is pertinent to other substances with a negative chronotropic effect. Such functional inhomogeneity is lacking in biological pacemakers.

1 Introduction

Whether or not there will be a place for *biological pacemakers* in addition to or as a substitute for technically very effective pacemakers, is discussed in another paper in this issue [53]. The biological material for *biological pacemakers* is derived from embryonic stem cells or from other types of cells with an undifferentiated status, or, alternatively, stems from geneticengineering techniques. Therefore, it is relevant to consider the pacemaker characteristics of embryonic hearts and myocytes.

2 The Embryonic Heart

2.1 The Initiation of the Heart Rhythm: Early Sinus Node Control

The embryonic heart develops by the fusion of two primordia into one single tube [60]. This tube bends and septates, eventually leading to the four-chambered heart [8].

J.A.E. Spaan et al. (Eds.): Biopacemaking, BIOMED, pp. 6–26, 2007.
springerlink.com

Fano and Badano [21] were the first in 1890 to show that there is a pacemaker hierarchy within the embryonic heart. They observed the propagation of contraction waves in longitudinal strips cut from the chicken heart, but they also noticed the higher contraction rate of the atrium compared with the ventricle after transection between these two segments. However, at an earlier stage of development Sabin [54] had shown in 1917 that the onset of contraction is in the ventricle not in the atrium:

''It is interesting to note that there is no movement whatever in the vein, the entire twitching being confined to the ventricle proper.''

The onset of cardiac contraction in the chicken embryo is at 29–30 h of development, whereas the circulation of the blood starts at 38–40 h [46]. Patten and Kramer [46] noted in 1933 about the ventricular onset of contraction:

''The first contractions were not rhythmic in their recurrence. They appeared more as sporadic flutters of restlessness in the developing myocardium, manifested first in one area and then in another, but always limited to the ventricle. (...) If one watches a heart at this stage for an hour or more, making repeated counts of the beats and recording the rest periods, it becomes apparent that the series of beats are tending to become longer and the rest intervals shorter. In other words a definite rhythmicity is gradually becoming established.''

The latter sentence from this famous paper of Patten and Kramer [46] confuses contractility with automaticity. Sabin [54] and Patten and Kramer [46] optically observed the onset of excitation-contraction coupling rather than the onset of automaticity. It was thought that there was a gradual development of pacemaker hierarchy from ventricle towards the atrium, ultimately ending in pacemaker dominance in the sinus venosus [3]. Van Mierop [59] has shown that this point of view is wrong. Early electrophysiological measurements were made by Meda and Ferroni [34] in 1959, showing the presence of diastolic depolarization in the sinus venosus and its absence in the ventricle in embryonic chicken hearts of 42 h of development. Van Mierop [59] explicitly made clear that there is electrical activation in hearts before they start to contract. Thus, he impaled hearts from 28 h of incubation and observed that there was electrical propagation of action potentials, originating from the sinus venosus, towards the ventricle even in completely noncontracting embryonic hearts. At a slightly later stage of development he noted:

''In 31 h embryos (...) the bulboventricular part of the heart was always seen to beat, the sinoatrial part never. Here again, however, sinoatrial action potentials could be recorded which were followed after an interval of about 100 ms by a ventricular action potential and a peristaltoid ventricular contraction.''

Obviously, the early observations on contractility have obscured the electrical pacemaker behaviour of the early embryonic heart for decades. The fact that the sinus node controls heart rate in very early embryonic stages, does not mean that the embryonic ventricle would not display automaticity if it were not paced.

This auxiliary characteristic relies on the fact that embryonic ventricle, at least in the mouse, has a sinus node type of pacemaker current based on HCN4 channels, which disappear a couple of days before birth [66] (see below).

Although it is clear that pacemaker dominance resides within the posterior side of the tubular heart (i.e. at the venous input side), it remains enigmatic how the sinus node develops towards its morphological recognizable structure, to quote Moorman and colleagues [37]. Recent advances in molecular biology have made clear that the transcription factor Tbx3 is unique for the central part of the cardiac conduction system, which comprises the sinus node, the atrioventricular node, the atrioventricular bundle and the proximal part of the bundle branches [25]. Where Tbx3 is expressed, chamber type molecular markers as connexin40 or natriuretic precursor peptide A are absent [25].

2.2 Embryonic Cardiac Development After the Initiation of a Regular Heart Rhythm

Until 6 days of development, that is after the establishment of regular sinus node function, the embryonic heart functions (1) without an AV node, (2) without a specific conduction system and (3) without valves. The origin of the AV node and of the ventricular conduction system has long been a matter of controversy (for older literature, see [7]). The AV node develops from the lower part of the interatrial septum and not from the AV canal, i.e. the zone between the atrium and the ventricle in the tubular heart, which becomes the *annulus fibrosis* in the adult heart [2]. Even after the development of the AV node (at about 6 days of development) there is no electrical communication between the atrium and the ventricle by this pathway, simply because the AV node and the ventricular septum are not yet connected [2]. The ventricular conduction system develops in situ, from the ventricular trabeculae [60]. In the embryonic heart at 1.5 days of development there is already some electrophysiological differentiation. The action potentials in (part of) the atrium and the ventricle are different [34, 59]. At this stage there is coexpression of adult atrial and ventricular isomyosins in the entire tubular heart except in a part of the sinoatrial region [14]. Until 2 days of development there is no zone in the heart in which the conduction velocity exceeds 2 cm/s [1]. Thus, there is slow conduction in the whole tubular heart. The conduction is also slow compared with the central zone of the adult sinoatrial node of several mammalian species [42]. During further development (stages older than 14 according to Hamburger and Hamilton [23]) there is a gradual loss of the coexpression of both isomyosins in the atrium and the ventricle [55]. This correlates with further electrophysiological differentiation within the heart [1]. Thus, areas with synchronous contraction (faster conduction) tend to show single isomyosin expression, whereas areas with peristaltoid contraction (slow conduction) maintain the coexpression of both types of isomyosins [14]. The outflow tract of a developing embryonic heart at day 4 shows remarkably slow conduction. Figure 1 (from [15]) shows a unique feature of the embryonic heart. Traces 1 and 2 are unipolar electrograms recorded from the ventricle and the outflow tract respectively. This embryonic heart was paced from the left atrium (open triangle in trace 1). The ventricular depolarization is indicated with the open circle and the ventricular repolarization with the filled square. The arrow in trace 2 indicates the activation of the outflow tract. Finally, trace 3 shows a transmembrane potential from the outflow tract. The upstroke of the (slow) action potential coincides with the activation of the outflow tract in trace 2. Also, the stimulation artefact of the next cycle in the left atrium occurs in the middle of the plateau phase of the action potential in the outflow tract (indicated with the large filled triangle in trace 3). The role of this very late activating (and contracting) area is probably to provide the embryonic heart with a

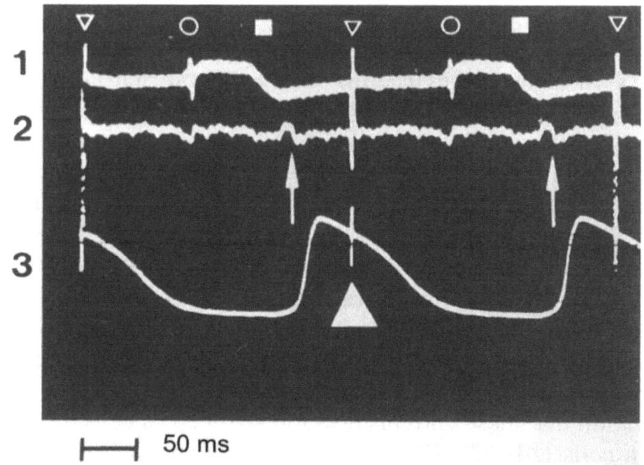

Fig. 1. Electrical recordings from embryonic chicken heart of 4 days old. *Traces 1* and *2* unipolar electrograms from ventricle and outflow tract. *Trace 1* unipolar electrogram from ventricle. *Open triangle* stimulus artefact from pacing at left atrium. *Open circle* activation of ventricle. *Filled square* repolarization of ventricle. *Trace 2* unipolar electrogram from outflow tract. *Arrow* repolarization of outflow tract. *Trace 3* transmembrane potential from outflow tract. The stimulation artefact of the next cycle in the left atrium (*trace 1*) occurs in the mid of the plateau phase of the action potential in the outflow tract (indicated with the *large filled triangle in trace 3*). Reproduced with permission from [15].

physiological substitute for valves that have not been developed at this stage. In the adult heart this area becomes the right ventricular outflow tract. It is remarkable that this area maintains the feature of final latest activation as has been described in many species under several circumstances, but recently also in the human heart under complete physiological conditions [51]. Apart from the fact that the outflow tract constitutes an interesting feature of cardiac development, it is emphasized (although completely speculative), that abnormal remnants of such embryological zones may play a role in syndromes and diseases as right ventricular dysplasia or Brugada syndrome, which are thought to have primarily a molecular background during recent years. The debate of the underlying mechanism behind Brugada syndrome has recently focused on the significance of conduction slowing [10] versus repolarization disorders [35] with or without a prominent role for so called channelopathies, in this case a mutation in the gene encoding the fast inward sodium channel. It is quite remarkable that the large majority of patients in a syndrome with supposed genetic background do in fact not carry the mutation [11]. It might be that a genetic disorder is not the cause of the syndrome, but rather a modulating factor. The reason for elaborating on this issue is that it might indicate that introducing tissue with inherent slow conduction, derived from embryonic tissue or mimicking characteristics of embryonic tissue, like a biological pacemaker, may come with the price of proarrhythmia.

2.3 Sequential (Ontogenetic) Appearance of Membrane Currents

Recent progress in genetic engineering has renewed the interest in the early development of the mouse heart. The order by which membrane currents appear in the embryonic heart has thus far been restricted to studies in mice embryos [12] and in cell cultures derived from mouse embryonic stem cells (for review see [24]). The slow inward calcium current ($I_{Ca\text{–}L}$) has been demonstrated at 9.5 days post coitum (dpc) [30] and increases steadily until birth at 19 dpc, [12, 30], whereas the fast inward sodium current (I_{Na}) becomes prominent at later stage [12]. With respect to the repolarizing currents the transient outward current (I_{to1}) develops first [12, 24, 67] with higher atrial than ventricular density [12]. Other outward potassium currents develop later with different regional densities [12]. Thus, cells at 11–13 dpc depend on $I_{Ca\text{–}L}$ for the upstroke and on I_{to}1 for repolarization of their action potentials in line with the observation that these currents develop also first in cells derived from mouse embryonic stem cells [24, 67]. Figure 2 (from [67]) shows a putative scheme with sequential development of membrane currents in the embryonic murine ventricle. The horizontal arrow at the top of Fig. 2 shows supposed development, whereas the arrow at the bottom of Fig. 2 indicates the order by which membrane currents might (re)appear when the view is taken that under certain pathological conditions a foetal gene program is recapitulated. It goes without saying that the latter is highly speculative. The pacemaker current I_f plays a prominent role in pacemaking in the adult sinus node, also in man. Figure 2 shows that this current disappears around birth from the embryonic murine ventricle. Reintroducing it in ventricle therefore introduces a current that is normally absent in ventricle.

2.4 Role of Pacemaker Currents with Focus on I_f

The embryonic mouse heart starts to beat at 8.5 dpc. The full gestation period takes 21 days. Figure 3a (taken from [66]) shows that ventricular myocytes from hearts at 9.5 dpc exhibit spontaneous activity and Fig. 3c shows the presence of inward current activating upon hyperpolarization, which is a feature of the pacemaker current I_f. At 18 dpc, that is 3 days before birth, the action potential configuration has changed substantially, spontaneous activity has slowed down and has lost regularity (Fig. 3b). The I_f current has disappeared almost completely shortly before birth (Fig. 3d). The fact that these ventricular myocytes display automaticity at early development does not exclude that the sinus node drives the embryonic heart from the very onset of electrical activity as in the chicken heart [59], but experimental proof for this is lacking. The right panel of Fig. 3 shows that the principal ion channel subunit at 9.5 dpc is based on expression of HCN4, which is a member of the hyperpolarization-activated cyclic nucleotide-gated (HCN) family of genes [4] and which is underlying the I_f current of the adult sinus node. During the second half of embryonic development the expression of HCN4 mRNA disappears almost completely (Fig. 3, right panel). HCN2, which is virtually the only expressed HCN gene in adult working atrial and ventricular myocytes, displays low expression during the full period of embryonic development. The role of the HCN2 based I_f current in adult atrium and ventricle is unclear, given the negative potential range where this current activates [45]. It thus seems as if immature ventricular myocytes are more or less sinus node type cells and that during the process of maturation the loss of sinus node type I_f current and the gain of inward rectifier current (I_{K1}) [12, 67] results in the loss of intrinsic automaticity.

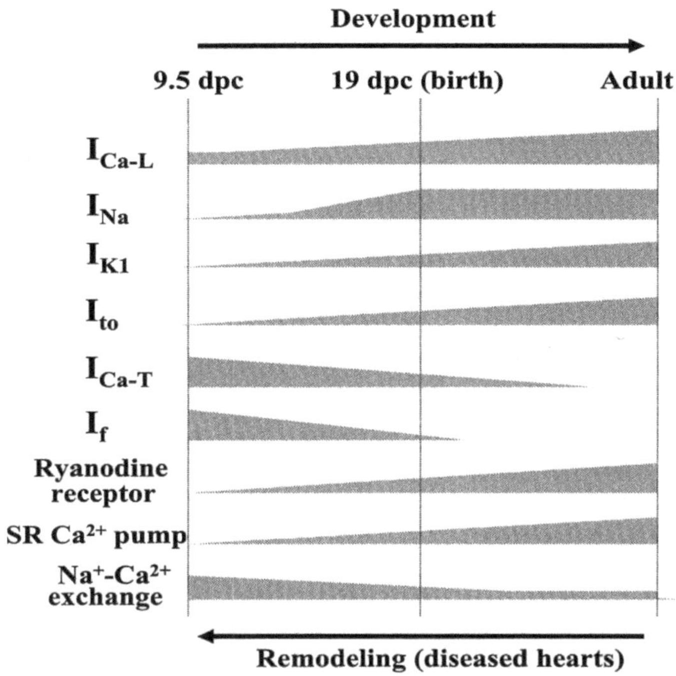

Fig. 2. Order by which membrane currents develop in the murine embryonic ventricle (*arrow at top*). During remodeling in pathophysiological processes these currents might reappear in reverse order (*arrow at bottom*). Reproduced with permission from [67].

2.5 ß-Adrenergic Modulation of Currents in Embryonic Myocytes

As early as at 9.5 dpc I_f current is responsive to ß-adrenergic modulation [66], as is the case with the L-type Ca^{2+}current [30]. The fact that I_f is more responsive to forskolin than to isoproterenol [66] suggests that the intracellular second messenger cascade develops earlier than (coupling to) the ß-adrenoceptor in the sarcolemma. Directly after birth the responsiveness to ß-adrenergic stimulation continues to increase as has been demonstrated in chicken [56].

Interestingly, in cardiomyocytes differentiated from human embryonic stem cells, positive chronotropic effects have been reported in response to phenylephrine (10^{-4} M), an α-adrenergic agonist, and to isoprenaline (10^{-6} M), a ß-adrenergic agonist. Also, negative effects of exposure to carbachol (10^{-4}M) were reported, which suggests effective vagal responsiveness [38]. However, the concentrations of all these substances were very high.

3 Pacemaking in the Adult Sinus Node

3.1 Regularity and Basic Cycle Length

Soon after the development of the technique to isolate individual myocytes from a whole heart [49], it was refined to the sinus node [33]. It came much as a surprise that single sinus node cells, despite the fact that they have been isolated from an intact adult sinus node, no longer possess the feature of beat-to-beat regularity (Fig. 4, taken from [43]). Also, an isolated sinus node, detached from the right atrium beats faster than when attached to the right atrium [27] and nodal areas close to the crista terminalis have a higher intrinsic frequency than the pacemaker area, which appears 'leading' in an intact sinus node [28]. The latter results from a steeper diastolic depolarization in combination with shorter action potential duration [43]. These intrinsic characteristics change when the nodal cells are interconnected with each other and with the surrounding, hyperpolarizing, atrium. The function of the intact sinus node has been regarded as a process of synchronization rather than as process of conduction with a leading, dominant pacemaker delivering current to its surroundings [36]. Obviously, knowledge of the behaviour of single sinus nodal cells is insufficient to understand the behaviour of the whole sinus node. This issue will reappear when chronotropic effects and autonomic modulation are discussed (see below).

3.2 Mechanism of Pacemaking

The main distinction between sinus node cells and working myocardial cells, irrespective whether it concerns atrial or ventricular cells, is that the former do not possess physiologically relevant I_{K1} current. This feature is also seen in embryological cells at early stages [12, 24, 67]. This explains why sinus node cells do not have a resting membrane potential, do not have a threshold potential for excitation and exert automaticity. In fact, automaticity results from the discrepancy between the kinetics of early inward and outward currents when these are not cancelled by the over whelming conductance of the inward rectifier current (I_{K1}). The lack of relevant I_{K1} current also explains the theoretical resting potential of –38 mV, where a sinus nodal cell would become quiescent, which is more a compromise between the Nernst potentials for K^+ (about –90 mV) and Na^+(+70 mV) than in any other part of the heart. The fact that the maximum diastolic potential is much more negative (at about –65 mV) follows from the fact that the oscillations of the three important pacemaker mechanisms are more or less out of phase. The (outward) delayed rectifier current, probably primarily its rapid component (I_{Kr}), at least in man [61], drives the maximum negative potential of sinus node cell to an area where the other two currents (the pacemaker current I_f and the L-type Ca^{2+}current) can become operational. Which of these currents is the most critical for automaticity is not a very relevant physiological question, although it has given rise to ample debate [16, 26, 44]. It should be noted, however, that full blockade of either I_{Kr} [62] or the L-type Ca^{2+} current [64] is incompatible with pacemaking. Full blockade of the 'pacemaker current' (I_f) decreases heart rate, but does not prevent automaticity. The background,

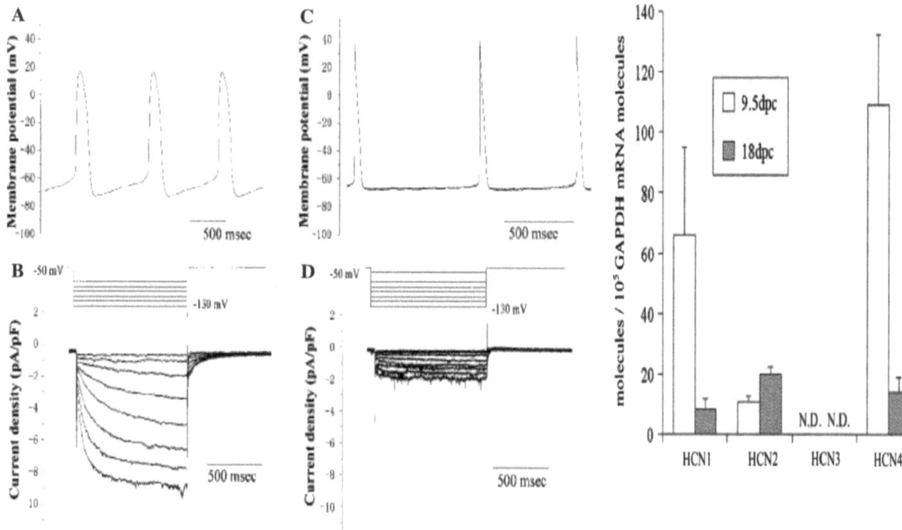

Fig. 3. Pacemaker current I_f in embryonic murine ventricle. Panel **a** action potentials at stage 9.5 days post coitum (*dpc*). Panel **b** action potentials at stage 18 dpc. Panel **c** activation of I_f upon hyperpolarizing pulses at 9.5 dpc. Panel **d** activation of I_f upon hyperpolarizing pulses at 18 dpc. *Right panel* Expression of HCN1 mRNA, HCN2 mRNA and HCN4 mRNA at 9.5 and 18 dpc. Compiled from [66].

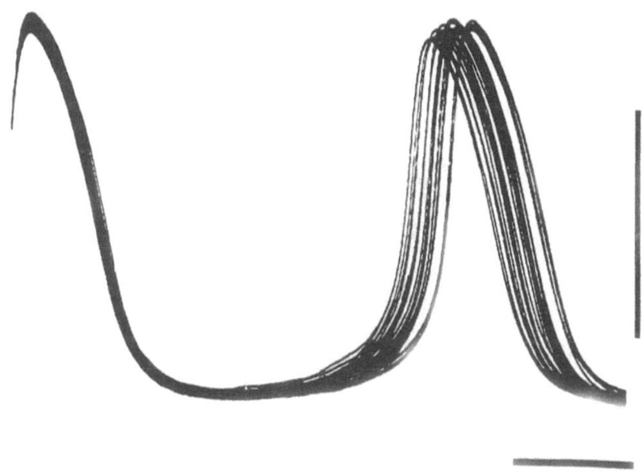

Fig. 4. Action potentials of a single sinus node cell isolated from a rabbit sinus node. There is no beat-to-beat regularity. *Horizontal bar* 100 ms. *Vertical bar* 50 mV. Reproduced with permission from [43].

however, is that blockade of I_{Kr} prevents the generation of a sufficiently negative membrane potential, which prevents the L-type Ca^{2+} current or I_f to become activated and that blockade of L-type Ca^{2+} current prevents the upstroke of action potentials in sinus node cells. Blockade of the I_f current, either by Cs^+or by alinidine or comparable agents, affects peripheral nodal cells more than the primary pacemaker area and thus does not disturb the standard nodal activation pattern [41].

4 Autonomic Modulation of the Adult Sinus Node

The pacemaker current involved in pacemaking are all more or less sensitive to environmental changes, amongst which autonomic modulation. Figure 5 (taken from [6]) shows a scheme of the rabbit sinus node with the dominant, leading (or primary)

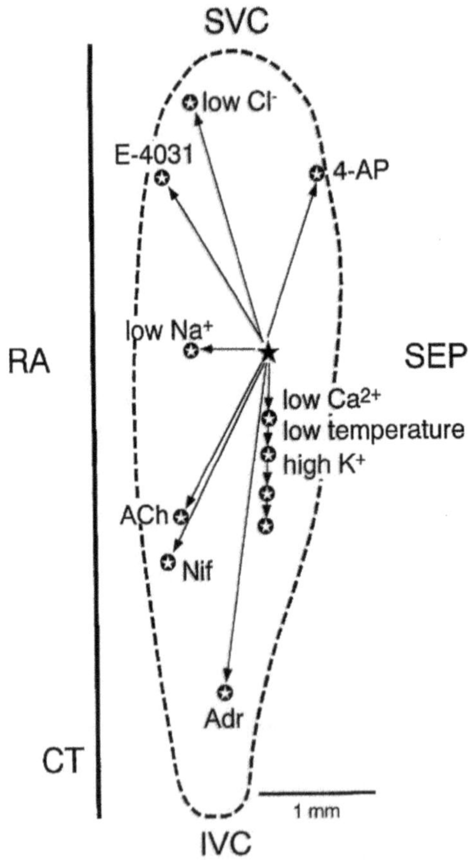

Fig. 5. Pacemaker shifts in response to several interventions. *CT* crista terminalis, *SEP* interatrial septum, *SVC* vena cava superior, *IVC* vena cava inferior, *E-4031* blocker of rapid component of delayed rectifier current, *4-AP* 4-amino pyridine (blocker transient outward current). *Adr* adrenaline, *Ach* acetylcholine, *Nif* nifedipine (blocker L-type Ca^{2+} current) Reproduced with permission from [6].

pacemaker area at the asterisk. Numerous changes in either concentrations of ions, or in autonomic tone or in circulating neurohumoural factors or blockers of specific membrane currents all may give rise to different local responses and therefore change the activation pattern of the nodal area. Because of the different responsiveness of those areas the chronotropic response of the intact sinus node cannot easily be predicted from the chronotropic responses of its constituent cells. The sinus node comprises several morphological cell types [63]. In addition there are also functional differences between more central (typical) nodal cells and peripheral (latent) pacemaker cells [5], which never the less have a higher intrinsic pacemaker potency [43]. Figure 6 (taken from [39]) shows the result of a computer analysis of cell types in the rabbit sinus node by discriminant analysis. Based on a combination of morphological and electrophysiological characteristics a specific type was assigned to each sinus node cell of which the location was unknown to the computer. Figure 6 shows that typical nodal cells are surrounded towards the crista terminalis (CT) by transitional cells with normal excitability and towards the interatrial septum by transitional cells with low excitability.

This is called *functional inhomogeneity* and we will explain this concept in more detail in the next section, because it is highly relevant for the requirements of *autonomic modulation* of a *biological pacemaker*.

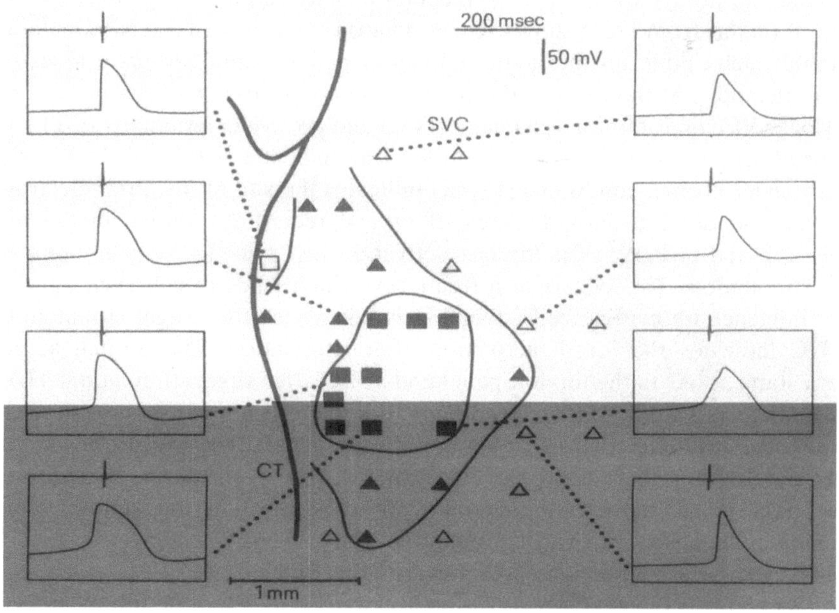

Fig. 6. Computer analysis of different sinus node cell types based on morphological and electrophysiological characteristics. *Filled squares* typical nodal cells. *Filled triangles* transitional cells with normal excitability. *Open triangles* transitional cells with low excitability. *Open square* atrial cell. The computer assigned the cell type by discriminant analysis, but was not aware of the location of the cells. Reproduced with permission from [39].

4.1 Functional Inhomogeneity

Figure 7 (taken from [32]) shows the background of functional inhomogeneity and its consequences for nodal chronotropic effects. Intact rabbit sinus nodes were sequentially impaled and activation patterns were determined during standard conditions (no adrenaline, no acetylcholine) and during the presence of either adrenaline (Adr) or acetylcholine (Ach). This gave rise to three different pacemaker centres (see also Fig. 5), one located in the superior sinus node (S, neither Adr nor Ach present), one located in the inferior sinus node (I, in the presence of Adr) and one located in the transitional zone (Tr), closer to the crista terminalis (CT, in the presence of Ach). These three centres were separated from each other. This will deliver three preparations ('S', 'I' and 'Tr') from each individual sinus node. Next the chronotropic responses to adrenaline and acetylcholine were determined for each of these centres. The responses of the primary centre (S) were intermediate both to acetylcholine and to adrenaline. The Adr centre (I) had large responses to both (neuro)humoural factors. The chronotropic responses to acetylcholine are depicted by dashed lines at the top of the histogram. There were individual preparations that turned quiescent in response to acetylcholine. In contrast, the Ach centre (Tr) hardly changed its cycle length after administration of either substance. The functional significance of these data is important. It indicates that there is huge intranodal variability with respect to receptor density and probably also innervation. Figure 8 (taken from [22]) shows recent data on the effect of vagal stimulation of the rabbit sinus node on the nodal activation pattern simultaneously assessed by optical methods. At the top left the field of view is shown with the orifices of the superior (SVC) and inferior vena cava (IVC) and the crista terminalis (CT) and the interatrial septum (IAS). The black dotted line indicates a line of block which is present under normal conditions [5]. EG indicates the site of the atrial electrogram. The black square indicates the area of optical recordings shown in the bottom panels a, b and c. Panel A is the last activation just prior to postganglionic vagal nerve stimulation. This occurs at a frequency which leads to neural firing with no direct influence on cardiac cells. Panel B is taken just after vagal stimulation and panel C indicates the fourth activation after stimulation. The normal activation pattern starts at 'A' in the top left panel and at 'B' after vagal stimulation. The shift is immediate and leads to a changed activation pattern for four consecutive cycles (see also the atrial electrogram in the upper right panel). The white zones in panels B and C show that there are areas that turn electrically quiescent. In addition, the upper right panel shows a hyperpolarization by 16% in the centre, which is dominant in the absence of vagal stimulation (trace 1).

This hyperpolarization increases towards the block zone (33%; trace 4). The importance of this recent study of Fedorov et al. [22] is that it definitely shows that vagal stimulation can turn areas of the sinus node inexcitable without complete depression of pacemaker function of the complete sinus node. The lesson to be learned here is that a biological pacemaker with strong *homogeneous* response to vagal stimulation or to comparable stimuli may not constitute a sound goal.

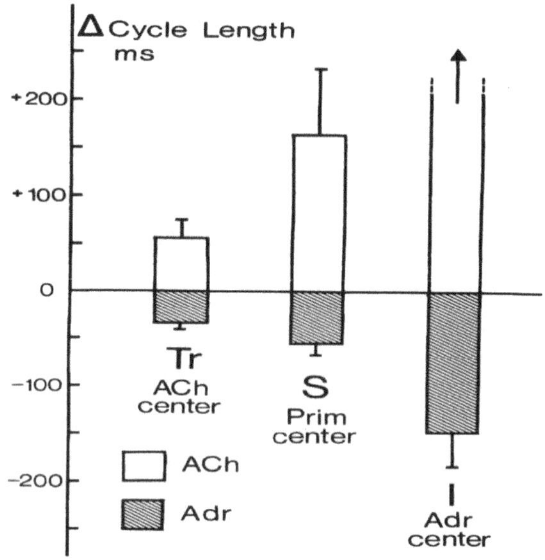

Fig. 7. Intact sinus nodes were mapped in normal Tyrode and in the presence of adrenaline (*Adr*; 0.6 μM) or acetylcholine (*Ach*; 5.5 μM). Subsequently these three centres (*S*: superior (primary) centre (Prim); *I*: inferior centre leading in the presence of Adr; *Tr*: transitional centre leading in the presence of Ach) were separated from each other and their chronotropic responses to Adr and Ach were measured. Such responses of subgroups are masked in the intact sinus node by pacemaker shifts. This phenomenon is known as *functional inhomogeneity*. Note that the inferior centre shows large responses to both Adr and Ach. The response to the latter may result in quiescence (indicated by *dashed lines at the top of the bar*). Reproduced with permission from [32].

It should be noted that agents and/or neurohumoural factors with a positive chronotropic effect select for intranodal sites with a high intrinsic responsiveness. Agents with a negative chronotropic effect, however, select for intranodal sites with a low intrinsic responsiveness. As a consequence, an intact sinus node will always display composite chronotropic responses because pacemaker shifts will obscure the responses of specific intranodal sites. The Ach (Tr) centre shows an important feature: it hardly responds to acetylcholine and, thereby, prevents standstill of a normal sinus node. if a biological pacemaker were innervated in combination with a homogeneous distribution of muscarinic (M2) receptors, the risk of complete standstill is evident. This does not seem an attractive property of a biological pacemaker.

4.2 The Case of Accentuated Antagonism

The same amount of (nor)epinephrine produces more acceleration, if the acetylcholine concentration or vagal tone is higher [29, 31, 40]. This phenomenon has been named ‘accentuated antagonism’ and has been explained by interaction between the two limbs of the autonomic nervous system at the preand postjunctional sites [29, 69]. Although this explanation may seem valid, it should be noted that functional inhomogeneity provides an alternative and more simple explanation. A high vagal tone shifts pacemaker dominance to areas with low responsiveness to both transmitters, whereas a high sympathetic tone shifts pacemaker dominance to areas with high responsiveness to both transmitters (Figs. 5, 7).

In addition, it should be realized that autonomic modulation affects the steepness of diastolic depolarization. A doubling of the slope of diastolic depolarization has a much more prominent effect at low heart rate than at high heart rate. Therefore it is logical that vagal and sympathetic effects are cycle length dependent [40]. At the level of single sinus node cells Rocchetti et al. [52] have unequivocally demonstrated that time-domain measurements of cycle length variability (changes in heart rate variability) cannot be translated into changes in neural input: "any condition depressing diastolic depolarization rate (DDR) may enhance cycle length (CL) variability, independent of changes in the pattern of neural activity" [52]; see also [68].

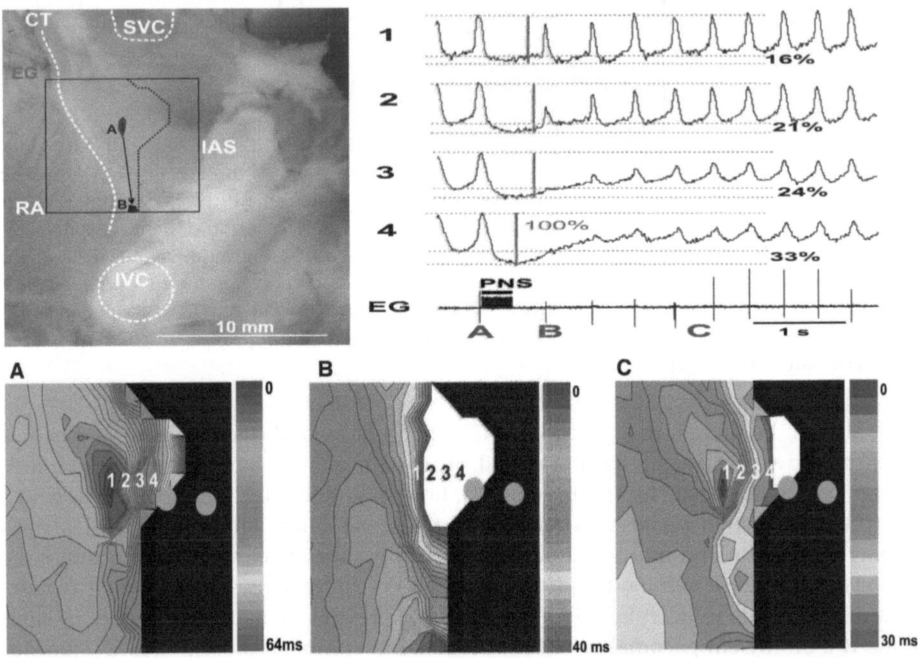

Fig. 8. Rabbit sinus node. *Top left* outline of preparation. *Black square* phrame of optical recordings as in *bottom panels* (**a**), (**b**) and (**c**). '*A*' and '*B*' indicate sites of pacemaker dominance during control and postganglionic vagal stimulation. EG site of electrogram in top *right panel*. *CT* crista terminalis; *IAS* interatrial septum. *SVC* superior vena cava; *IVC* inferior vena cava. *Bottom panels* activation patterns during control conditions (**a**), during the first cycle after postganglionic vagal stimulation (**b**) and during the fourth cycle after postganglionic vagal stimulation (**c**). Optical recordings of action potentials at the sites 1, 2, 3 and 4 in the bottom panels appear in the top right panel together with the amount of hyperpolarization of the maximum diastolic potential given as a percentage. See text for further details. Reproduced with permission from [22].

4.3 Differences Between Neural and Humoural Effects

It has been observed in dogs that bilateral vagal stimulation results in a lower heart rate in combination with stellate ganglion stimulation than in combination with norepinephrine infusion [57]. Although an intricate prejunctional interaction may explain this observation, functional inhomogeneity may provide a more simple explanation for this remarkable difference. High vagal tone will shift the pacemaker to the area with lowest innervation. Such an area may still accelerate in response to circulating catecholamines, but not or much less to sympathetic stimulation, simply because the nerves do not impinge on that particular area.

5 The Transplanted Heart

The transplanted heart is a rich source of information for the relation between the autonomic nervous system and the heart [9]. Blood pressure results from the product between cardiac output and peripheral resistance. Cardiac output is—in its turn—composed of the product between heart rate and stroke volume. The basic physiological concept that only cardiac output is directly, i.e. on a 'per cycle' basis is affected by vagal influences, whereas peripheral resistance is only affected by sympathetic influences with its concomitant 0.1 Hz dominant frequency seems untenable, given the fact that recipients of a donor heart are able to stand up at all. However, this reasoning applies in fact also to a normal heart.

The pivotal loss of information after heart transplantation is not the loss of information from the receptors in the sinus caroticus and in the arch of the aorta (baroreflex), but from the intracardiac receptors. Indeed, the main problems of these patients are blunted responses to volume expansion and to natriuremia. Thus these patients function in a permanent state of a large circulating volume and hypertension [9].

It has been reported that—at least—sympathetic reinnervation may occur after cardiac transplantation [65]. The evidence is based on the fact that the donor heart is able to produce noradrenaline after an injection with tyramine in the left anterior descending or circumflex coronary arteries. Although very interesting and probably functionally relevant, such experiments prove that the nerves in the donor heart still have or have regained metabolic activity. In my opinion it does not unequivocally prove that the central nervous system of the recipient has sympathetic neural control over the donor heart. In dogs with autotransplantation it has been reported that the maximal heart rate during exercise increases much more slowly and to a substantially lower maximum heart rate [17]. Never theless, in patients with a transplanted heart the maximum exercise tolerance can be large [9].

By and large, the most severe problems in patients with a transplanted heart, besides the problems with rejection, reside in the relation between heart and kidneys and focus on regulation of the total blood volume and blood pressure and to a lesser extent on exercise tolerance and postural changes. Probably this results from the fact that the intrinsic response to preload changes of donor hearts (Starling law) remains—at least partially—effective. I think, therefore, that innervation of a *biological pacemaker* is less important than the capacity to respond to circulating catecholamines (see below).

6 Biological Pacemakers

6.1 State of the Art

Thus far moderate success has been achieved with biological pacemakers either along the lines of a genetic or a cellular approach (Rosen et al., this issue [53]). Successful introduction of human ß2-adrenoceptor constructs has been reported in the murine heart [19] and in the porcine heart [20]. As far as the in vivo parts of these studies are involved, I wish to underscore that demonstrating that heart rate can be increased after the introduction of components of the adrenergic system is not the same as demonstrating that the heart is able to respond to catecholamines. Secondly, the response to injection of these constructs persisted for 2–3 days in the mice [19]) and for less than 2 days in the pigs [20]. Obviously, these responses are transient. In the case of cellular approaches it is important to know what happens to the implanted or injected biomaterials in case of loss of function. The biological function is simply lost, when the implant disappears (is 'eaten'), but when it is still sitting in the myocardium and has lost pacemaker potency, a proarrhythmic risk may ensue.

In the dog substantial success has been achieved thus far (Rosen et al., this issue [53]) both by a gene therapy approach [47, 50] and by a cellular approach based on adult human mesenchymal stem cells [48]. In the former approach injection of adenoviral contructs with mouse HCN2 constructs into the left canine atrium [47] yielded spontaneous rhythms during vagal stimulation (in order to silence the native sinus node). This occurred 3–4 days after injection. Left atrial myocytes isolated from these hearts showed prominent I_f current. In a comparable study HCN2 adenovirus contructs were injected in the posterior limb of the left bundle branch of canine hearts [50]. Again, during vagal stimulation ventricular escape rhythms were demonstrated at least 7 days after the injection. There was a brief period of arrhythmias after the injection, but this appears to be related to the injection not to the construct, because the arrhythmias were also prominent in the control group and ceased after days. Although these results are encouraging, it is emphasized that positive chronotropy in response to catecholamines was not demonstrated. Along the cellular approach human mesenchymal stem cells transfected with the murine HCN2 gene were injected in the epicardium of the left canine ventricle [48]. Again, during sinus arrest pacemaker activity was observed. In the in vitro part of this study [48], acetylcholine did not affect I_f current, although it could mitigate the response to isoproterenol. Although it had been reported that human mesenchymal stem cells can form functional gap junctions with freshly isolated canine ventricular myocytes [58], this was also demonstrated in vivo at the actual site of injection of the engineered mesenchymal stem cells [48].

An experimental proof for the putative scheme as shown in Fig. 7 of Rosen et al. [53] (this issue), in which a genetically engineered stem cell is able to deliver pacemaker current to a myocyte is provided in another paper in this issue [13]. Thus, the spontaneous beating rate of neonatal rat cardiomyocytes can be tuned by coculturing them with HEK–293 cells which stably express murine inward rectifier channels (Fig. 3 in [13], this issue). Although the beating rate of the neonatal rat cardiomyocytes rapidly decreases with a small proportion (only 5%) of engineered HEK-293 cells, spontaneous beating did not cease even when the large majority of cells (75%) were HEK-293 cells. With full blockade of I_{K1} current, more subtle tuning of the remaining ensemble spontaneous beating rate resulted. Although the authors did not emphasize this themselves, it is of particular interest that even with full blockade of I_{K1} current, the percentage of HEK-293 cells remains important for the final beating rate. This resembles the natural situation in which an intact, but isolated sinus node has a higher frequency than a complete right atrium, where the sinus node is connected to the surrounding atrial muscle [27].

6.2 Requirements of a Biological Pacemaker

I cite here a phrase from the paper by Rosen et al. [53] elsewhere in this issue: ''...we have taken a lesson from our engineering colleagues who designed the electronic pacemaker; that is we are working to fine-tune a structure that mimics the sinus node functionally without recapitulating it morphologically.'' Creating a homogeneous biological sinus node with functional responses to neurotransmitters with a prospect to real innervation, or at least with adrenergic-and muscarinic-type responses to humoural factors seems at first glance a desirable goal, but might probably not be such a good idea. The reason is simple. Such a biological pacemaker would also have the capacity of quiescence. This is exactly what we do not want. The ideal biological pacemaker would be one that is able to cope with postural changes and exercise. The former goal may seem far-fetched. It requires innervation, because adaptations are needed within a single cardiac cycle. For a response to exercise it is sufficient that a biological pacemaker can increase its rate. The biological pacemaker *in statu nascendi* as proposed by Rosen et al. [53] in this issue fulfils this more moderate goal and has a limitation that may constitute two advantages that can become important in future competition with the electronic pacemaker. The biological pacemaker is based on the pacemaker current (I_f) only, not on a combination of multiple membrane currents, e.g. the acetylcholine sensitive K^+current ($I_{K\text{-}Ach}$) is lacking. Therefore it cannot easily turn quiescent. It can only accelerate, which is exactly what an electronic pacemaker cannot. The debate on the mechanism of vagal modulation of sinus rhythm has never been definitely settled. Either acetylcholine inhibits I_f current or it increases $I_{K\text{-}Ach}$ current or it does both (see for details Boyett et al. 2000). Figure 9 (taken from [18]) focuses on this issue. The left panel shows the effect of postganglionic vagal nerve stimulation (same technique as applied by Fedorov et al. [22]; see Fig. 8 in this paper), coupled to the cardiac cycle in the isolated right atrium of the rabbit. The top left panel shows the effect 10 stimuli per cycle leading to a prolongation of cycle length from 456 to 531 ms. The bottom left panel shows the same experiment in the presence of 3 μM atropine, blocking the vagally mediated response. During this procedure a hyperpolarizing pulse was given during diastole (right panel). This leads to an electrotonic disturbance of the membrane potential during diastole, which can be followed at a distance from the site of current injection. When the membrane conductance is high, much current 'escapes' over the sarcolemma and little current is transported along the axial pathway, the conductance of which is determined by cytoplasm and gap junctions. The right panel shows that during vagal stimulation the electrotonic potential decreases, which by simple Ohm's law means that the resistance of the sarcolemma has decreased by opening of a membrane channel, not by closing. This provides a strong argument for opening of

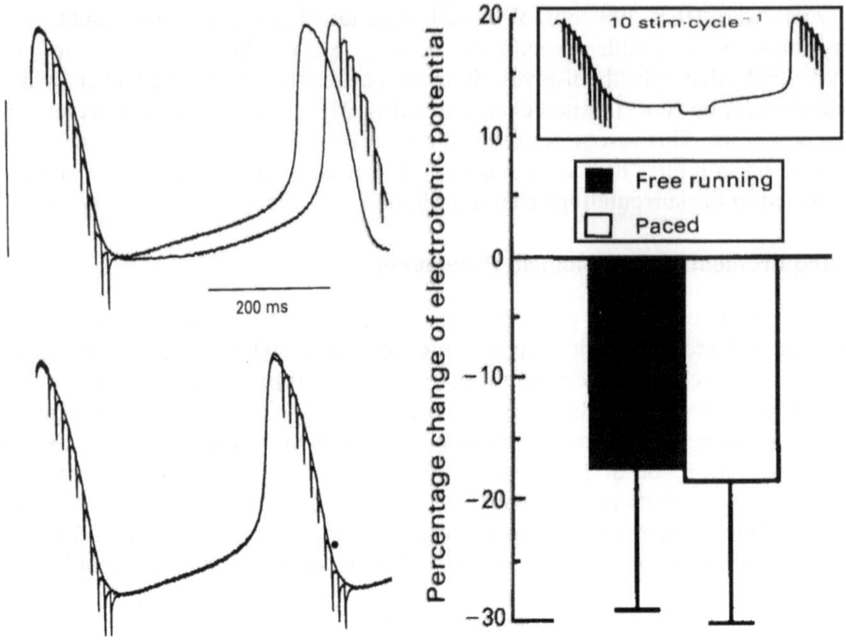

Fig. 9. Postganglionic vagal stimulation in the rabbit sinus node. The chrontropic effect of ten stimuli locked to the spontaneous cycle is shown in the *top panel left*. The *bottom panel left* shows the same response in the presence of 3 μM atropine. The *right panel* shows the electrotonic potential during diastole. There is a 15–20% decrease of this potential during vagal stimulation, which can only be explained when membrane resistance has decreased. This implies that vagal stimulation increases the overal membrane conductance. This is more compatible with an increase of $I_{K\text{-}Ach}$ than with a decrease of I_f in response to vagal stimulation. Compiled from [18].

$I_{K\text{-}Ach}$, not closing of I_f in response to physiologically relevant vagal stimulation. This supports my view that it is fortunate that the biological pacemaker of Rosen et al. [53] (this issue) is solely based on I_f current.

A relatively high intrinsic heart rate based on a biological pacemaker, without the possibility of deceleration (e.g. during rest or sleep) may also be useful from the point of view that cardiac muscle with suboptimal contractile performance has impaired capacitance to compensate low heart rate by a high stroke volume. For this reason the management of recipients of a transplanted heart aims at maintaining a relatively high resting heart rate, if possible above 100 beats/min [9].

7 Conclusion

Biological pacemakers have a long way to go before they will ever be superior to electronic pacemakers. There is, however, important progress. It seems sufficient when biological pacemakers aim at responsiveness to humoural factors not neural factors in order to be able to cope with exercise.

References

1. Argüello C, Alanis J, Pantoja O, Valenzuela B (1986) Electrophysiological and ultrastructural study of the atrioventricular canal during the development of the chick embryo. J Mol Cell Cardiol 18:499–510
2. Argüello C, Alanis J, Valenzuela B (1988) The early development of the atrioventricular node and bundle of His in the embryonic chick heart. An electrophysiological and morphological study. Development 102:623–637
3. Barry A (1942) The intrinsic pulsation rates of fragments of the embryonic chick heart. J Exp Zool 91:119–130
4. Biel M, Schneider A, Wahl C (2002) Cardiac HCN channels: structure, function, and modulation. Trends Cardiovasc Med 12:206–213
5. Bleeker WK, Mackaay AJC, Masson-Pe´vet M, Bouman LN, Becker AE (1980) Functional and morphological organization of the rabbit sinus node. Circ Res 46:11–22
6. Boyett MR, Honjo H, Kodama I (2000) The sinoatrial node, a heterogeneous pacemaker structure. Cardiovasc Res 47:658–687
7. Canale ED, Campbell GR, Smolich JJ, Campbell JH (1986) Cardiac muscle. Springer, Berlin
8. Christoffels VM, Burch JB, Moorman AFM (2004) Architectural plan for the heart: early patterning and delineation of the chambers and the nodes. Trends Cardiovasc Med 14: 301–307
9. Cirklin JK, Young JB, McGiffin DC (2002) Physiology of the transplanted heart. In: Cirklin JK, Young JB, McGiffin DC (eds) Heart transplantation. Churchill Livingstone, New York, pp 353–372
10. Coronel R, Casini S, Koopmann TT, Wilms-Schopman FJG, Verkerk AO, de Groot JR, Bhuiyan Z, Bezzina CR, Veldkamp MW, Linnenbank AC, van der Wal AC, Tan HL, Brugada P, Wilde AAM, de Bakker JMT (2005) Right ventricular fibrosis and conduction delay in a patient with clinical signs of Brugada syndrome: a combined electrophysiological, genetic, histopathologic, and computational study. Circulation 112:2769–2777
11. Coronel R, Berecki G, Opthof T (2006) Why the Brugada syndrome is not yet a disease: syndromes, diseases and genetical causality. Cardiovasc Res 72:361–363
12. Davies MP, An RH, Doevendans P, Kubalak S, Chien KR, Kass RS (1996) Developmental changes in ionic channel activity in the embryonic murine heart. Circ Res 78:15–25
13. De Boer TP, van Veen TAB, Houtman AJ, Jansen JA, van Amersfoorth SC, Doevendans PA, Vos MA, van der Heyden MAG (2007) Inhibition of cardiomyocyte automaticity by electrotonic application of inward rectifier current from Kir2.1 expressing cells. Med Biol Eng Comput (this issue). DOI 10.1007/s11517-006-0059-8
14. De Jong F, Geerts WJC, Lamers WH, Los JA, Moorman AFM (1987) Isomyosin expression patterns in tubular stages of chicken heart development: a 3-D immunohistochemical analysis. Anat Embryol 177:81–90
15. De Jong F, Opthof T, Wilde AAM, Janse MJ, Charles R, Lamers WH, Moorman AFM (1992) Persisting zones of slow impulse conduction in developing chicken hearts. Circ Res 71:240–250

16. DiFrancesco D (1993) Pacemaker mechanisms in cardiac tissue. Annu Rev Physiol 55:451–467
17. Donald DE, Shephard JT (1963) Response to exercise in dogs with cardiac denervation. Am J Physiol 205:494–500
18. Duivenvoorden JJ, Bouman LN, Opthof T, Bukauskas FF, Jongsma HJ (1992) Effect of transmural vagal stimulation on electrotonic current spread in the rabbit sinoatrial node. Cardiovasc Res 26:678–686
19. Edelberg JM, Aird WC, Rosenberg RD (1998) Enhancement of murine cardiac chronotropy by the molecular transfer of the human ß2 adrenergic receptor cDNA. J Clin Invest 101:337–343
20. Edelberg JM, Huang DT, Josephson ME, Rosenberg RD (2001) Molecular enhancement of procine cardiac chronotropy. Heart 86:559–562
21. Fano G, Badano F (1890) Etude physiologique des premiers stades dedévelopement du coeur embryonaire du poulet. Arch Ital Biol 13:387–422
22. Fedorov VV, Hucker WJ, Dobrzynski H, Rosenshtraukh LV, Efimov IR (2006) Postganglionic nerve stimulation induces temporal inhibition of excitability in the rabbit sinoatrial node. Am J Physiol Heart Circ Physiol 291:H612–H623
23. Hamburger V,Hamilton HL (1951) A series of normal stages in the development of the chick embryo. J Morphol 88:49–92
24. Hescheler J, Fleischman BK, Lentini S, Maltsev VA, Rohwedel J, Wobus AM, Addicks K (1997) Embryonic stem cells: a model to study structural and functional properties in cardiomyogenesis. Cardiovasc Res 36:149–162
25. Hoogaars WMH, Tessari A, Moorman AFM, De Boer PAJ, Hagoort J, Soufan AT, Campione M, Christoffels VM (2004) The transcriptional repressor Tbx3 delineates the developing central conduction system of the heart. Cardiovasc Res 62:489–499
26. Irisawa H, Brown HF, Giles W (1993) Cardiac pacemaking in the sinoatrial node. Physiol Rev 73:197–227
27. Kirchhof CJHJ, Bonke FIM, Allessie MA, Lammers WJ (1987) The influence of the atrial myocardium on impulse formation in the rabbit sinus node. Pflügers Arch 410:198– 203
28. Kodama I, Boyett MR (1985) Regional differences in the electrical activity of the rabbit sinus node. Pflügers Arch 404:214–226
29. Levy MN (1971) Sympathetic–parasympathetic interactions in the heart. Circ Res 29:437–445
30. Liu W, Yasui K, Arai A, Kamiya K, Cheng J, Kodama I, Toyama J (1999) ß-Adrenergic modulation of L-type Ca^{2+} channel currents in early stage embryonic mouse heart. Am J Physiol 276:H608–H613
31. Mackaay AJC, Opthof T, Bleeker WK, Jongsma HJ, Bouman LN (1980) Interaction of adrenaline and acetylcholine on cardiac pacemaker function. J Pharmacol Exp Ther 214:417–422
32. Mackaay AJC, Opthof T, Bleeker WK, Jongsma HJ, Bouman LN (1982) Interaction of adrenaline and acetylcholine on sinus node function. In: Bouman LN, Jongsma HJ (eds) Cardiac rate and rhythm. Martinus Nijhoff, The Hague, pp 507–523
33. Masson-Pévet M, Jongsma HJ, Bleeker WK, Tsjernina L, van Ginneken ACG, Treijtel BW, Bouman LN (1982) Intact isolated sinus node cells from the adult rabbit heart. J Mol Cell Cardiol 14:295–299
34. Meda E, Ferroni A (1959) Early functional differentiation of heart muscle cells. Experientia 15:427–428
35. Meregalli PG, Wilde AAM, Tan HL (2005) Pathophysio logical mechanisms of Brugada syndrome: depolarization disorder, repolarization disorder, or more? Cardiovasc Res 67: 367–378

36. Michaels DC, Matyas EP, Jalife J (1987) Mechanisms of sinoatrial pacemaker synchronization: a new hypothesis. Circ Res 61:704–714
37. Moorman AFM, De Jong F, Denyn MMFJ, Lamers WH (1998) Development of the cardiac conduction system. Circ Res 82:629–644
38. Mummery C, Ward-Van Oostwaard D, Doevendans P, Spijker R, Van Den Brink S, Hassink R, Van Der Heyden M, Opthof T, Pera M, Brutel De La Riviere A, Passier R, Tertoolen L (2003) Differentiation of human embryonic stem cells to cardiomyocytes. Role of coculture with visceral endoderm-like cells. Circulation 107:2733–2740
39. Opthof T,Bleeker WK, Masson Pevet M, Jongsma HJ, Bouman LN (1983a) Little-excitable transitional cells in the rabbit sinoatrial node: a statistical, morphological and electrophysiological study. Experientia 39:1099–1101
40. Opthof T, Mackaay AJC, Bleeker WK, Jongsma HJ, Bouman LN (1983b) Cycle length dependence of the chronotropic effects of adrenaline and acetylcholine in the rabbit sinoatrial node. J Autonom Nerv Syst 8:193–204
41. Opthof T, Duivenvoorden JJ, VanGinneken ACG, Jongsma HJ, Bouman LN (1986) Electrophysiological effects of alinidine (St 567) on sinoatrial node fibers in the rabbit heart. Cardiovasc Res 20:727–739
42. Opthof T, De Jonge B, Jongsma HJ, Bouman LN (1987a) Functional morphology of the mammalian sinuatrial node. Eur Heart J 8:1249–1259
43. Opthof T, Van Ginneken ACG, Bouman LN, Jongsma HJ (1987b) The intrinsic cycle length in small pieces isolated from the rabbit sinoatrial node. J Mol Cell Cardiol 19: 923–934
44. Opthof T (1988) The mammalian sinoatrial node. Cardiovasc Drugs Ther 1:573–597
45. Opthof T (1998) The membrane current (I_f) in human atrial cells. Implications for atrial arrhythmias. Cardiovasc Res 38:537–540
46. Patten BM, Kramer TC (1933) The initiation of contraction in the embryonic chick heart. Am J Anat 53:349–375
47. Plotnikov AN, Sosunov EA, Qu J, Shalpakova IN, Anyukhovsky EP, Liu L, Janse MJ, Brink PR, Cohen IS, Robinson RB, Danilo P, Rosen MR (2004) Biological pacemaker implanted in canine left bundle branch provides ventricular escape rhythms that have physiologically acceptable rates. Circulation 109:506–512
48. Potapova I, Plotnikov A, Lu Z, Danilo P, Valiunas V, Qu J, Doronin S, Zuckerman J, Shalapakova IN, Gao J, Pan Z, Herron AJ, Robinson RB, Brink PR, Rosen MR, Cohen IS (2004) Human mesenchymal stem cells as a gene delivery system to create cardiac pacemakers. Circ Res 94:952–959
49. Powell T, Twist VW (1976) A rapid technique for the isolation and purification of adult cardiac muscle cells having respiratory control and a tolerance to calcium. Biochem Biophys Res Comm 72:327–333
50. Qu J, Plotnikov AN, Danilo P, Shlapakova I, Cohen IS, Robinson RB, Rosen MR (2003) Expression and function of a biological pacemaker in canine heart. Circulation 107:1106–1109
51. Ramanathan C, Jia P, Ghanem R, Ryu K, Rudy Y (2006) Activation and repolarization of the normal human heart under complete physiological conditions. Proc Natl Acad Sci USA 103:6309–6314
52. Rocchetti M, Malfatto G, Lombardi F, Zaza A (2000). Role of the input/output relation of sinoatrial myocytes in cholinergic modulation of heart rate variability. J Cardiovasc Electrophysiol 11:522–530
53. Rosen MR, Brink PR, Cohen IS, Robinson RB (2007) Biological pacemakers based on I_f. Med Biol Eng Comput (this issue)

54. Sabin FR (1917) Origin and development of the primitive vessels of the chick and the pig. Carnegie Cont Embryol 6:61–124
55. Sanders E, De Groot IJM, Geerts WJC, De Jong F, Van Horssen AA, Los JA, Moorman AFM (1986) The local expression of adult chicken heart myosins during development. Anat Embryol 174:187–193
56. Satoh H, Sperelakis N (1993) Hyperpolarization activated inward current in embryonic chick cardiac myocytes: developmental changes and modulation by isoproterenol and carbachol. Eur J Pharmacol 240:283–290
57. Takahashi N, Zipes DP (1983) Vagal modulation of adrenergic effects of canine sinus and atrioventricular nodes. Am J Physiol 244:H775–H781
58. Valiunas V, Doronin S, Valiuniene L, Potapova I, Zuckerman J, Walcott B, Robinson RB, Rosen MR, Brink PR, Cohen IS (2004) Human mesenchymal stem cells make cardiac connexins and form functional gap junctions. J Physiol 555:617–626
59. Van Mierop LHS (1967) Location of pacemaker in chick embryo heart at the time of initiation of heartbeat. Am J Physiol 212:H407–H415
60. Van Mierop LHS (1979) Morphological development of the heart. In: Berne RM, Sperelakis N, Geiger SR (eds) Handbook of physiology, Sect. 2, vol 1. The heart. Williams and Wilkins Co., Baltimore, pp 1–28
61. Veldkamp MW, Van Ginneken ACG, Opthof T, Bouman LN (1995) Human delayed rectifier. Circulation 92:3497– 3504
62. Verheijck EE, van Ginneken ACG, Bourier J, Bouman LN (1995) Effects of delayed rectifier current blockade by E-4031 on impulse generation in single sinoatrial nodal myocytes of the rabbit. Circ Res 76:607–615
63. Verheijck EE, Wessels A, Van Ginneken ACG, Bourier J, Markman MWM, Vermeulen LJM, De Bakker JMT, Lamers WH, Opthof T, Bouman LN (1998) Distribution of atrial and nodal cells within the rabbit sinoatrial node. Models of sinoatrial transition. Circulation 97:1623–1631
64. Verheijck EE, van Ginneken ACG, Wilders R, Bouman LN (1999) Contribution of L-type Ca^{2+} current to electrical activity in sinoatrial nodal myocytes of rabbits. Am J Physiol 276:H1064–H1077
65. Wilson RF, Laxson DD, Christensen BV, McGinn AL, Kubo SH (1993) Regional differences in sympathetic reinnervation after human orthotopic cardiac transplantation. Circulation 88:165–171
66. Yasui K, Liu W, Opthof T, Kada K, Lee J K, Kamiya K, Kodama I (2001) The I_f current and spontaneous activity in mouse embryonic ventricular myocytes. Circ Res 88:536–542
67. Yasui K, Niwa N, Takemura H, Opthof T, Muto T, Horiba M, Shimizu A, Lee JK, Honjo H, Kamiya K, Kodama I (2005) Pathophysiological significance of T-type calcium channels: expression of T-type Ca(2+) channels in fetal and diseased heart. J Pharmacol Sci 99:205–210
68. Zaza A, Lombardi F (2001) Autonomic indexes based on the analysis of heart rate variability: a view from the sinus node. Cardiovasc Res 50:434–442
69. Zipes DP, Miyazaki T (1990) The autonomic nervous system and the heart: basis for understanding interactions and effects on arrhythmia development. In: Zipes DP, Jalife J (eds) Cardiac electrophysiology. From cell to bedside. W.B. Saunders, Philadelphia, pp 312–330

Creation of a Biological Pacemaker by Gene- or Cell-Based Approaches

Eduardo Marbán and Hee Cheol Cho

Institute of Molecular Cardiobiology, Division of Cardiology,
Johns Hopkins University School of Medicine, 858 Ross Bldg,
720 Rutland Ave, Baltimore, MD 21205, USA
marban@jhmi.edu

Abstract. Cardiac rhythm-associated disorders are caused by malfunctions of impulse generation and conduction. Present therapies for the impulse generation span a wide array of approaches but remain largely palliative. The progress in the understanding of the biology of the diseases with related biological tools beckons for new approaches to provide better alternatives to the present routine. Here, we review the current state of the art in gene and cell-based approaches to correct cardiac rhythm disturbances. These include genetic suppression of an ionic current, stem cell therapies, adult somatic cell-fusion approach, novel synthetic pacemaker channel, and creating a self-contained pacemaker activity in non-excitable cells. We then conclude by discussing advantages and disadvantages of the new possibilities.

1 Introduction

The heart requires a steady rhythm and rate in order to fulfill its physiological role as the pump for the circulation. An excessively rapid heart rate (tachycardia) allows insufficient time for the mechanical events of ventricular emptying and filling. Cardiac output drops, the lungs become congested, and, in the extreme, the circulation collapses. An equally morbid chain of events ensues if the heart beats too slowly (bradycardia). Serious disturbances of cardiac rhythm, known as arrhythmias, afflict more than three million Americans and account for >479,000 deaths annually [71]. In 2001, \$2.7 billion (\$6,634 per discharge) was paid to Medicare beneficiaries for cardiac arrhythmia-related diseases [71].

Current therapy has serious limitations: antiarrhythmic drugs can sometimes be effective, but their utility is limited by their propensity to create new arrhythmias while suppressing others [17, 19, 66, 74]. Ablation of targeted tissue can readily cure simple wiring errors, but is less effective in treating more complex and common arrhythmias, such as atrial fibrillation or ventricular tachycardia [20, 59]. Implantable devices can serve as surrogate pacemakers to sustain heart rate, or as defibrillators to treat excessively rapid rhythms. Such devices are expensive, and implantation involves a number of acute and chronic risks (pulmonary collapse, bacterial infection, lead or generator failure [6]). In short, arrhythmias are a serious threat of public health proportions, and current treatment is inadequate. Given these limitations, we have begun to develop gene or cell therapy as an alternative to conventional treatment.

J.A.E. Spaan et al. (Eds.): Biopacemaking, BIOMED, pp. 27–44, 2007.
springerlink.com

The most obvious application of gene therapy is to correct monogeneic deficiency disorders such as hemophilia or adenosine deaminase deficiency. Indeed, the latter is the only disease to have been cured (in a few infants) by gene therapy [7]. Gene therapy for cardiovascular disorders, as it is most commonly being developed today, focuses not on correcting deficiency disorders but rather on attempts to foster angiogenesis in ischemic myocardium [43, 63], or to suppress vascular stenosis in a variety of iatrogenic settings [44, 55]. The concept of gene or cell therapy for cardiac arrhythmias differs conceptually from conventional applications. We seek to achieve functional re-engineering of cardiac tissue, so as to alter a specific electrical property of the tissue in a salutary manner. For example, genes or cells are introduced to alter the velocity of electrical conduction in a defined region of the heart, or to create a spontaneously active biological pacemaker from normally quiescent myocardium. A relevant analogy is the use of off-the-shelf or customized parts to improve the performance of a lackluster automobile engine. Our ''parts'' are wild-type (or mutant) genes and engineered cells; our engine is the heart.

Here, we will review our progress in the creation of biological pacemakers. We then conclude by considering future directions of this type of gene therapy.

2 Biological Pacemaker by I_{K1} Knockout

The pacemaker of the heart is normally encompassed within a small region known as the sinoatrial (SA) node. The SA node initiates the heartbeat, sets the rate and rhythm of cardiac contraction, and thereby sustains the circulation [9]. The working muscle of the heart (myocardium), comprising the pumping chambers known as the atria and the ventricles, is normally excited by pacemaker activity originating in the SA node. However, in the absence of such activity, the rhythmic contraction and relaxation of myocardium discontinues. Therefore, loss of specialized pacemaker cells in the SA node, as occurs in a variety of common diseases, results in circulatory collapse, necessitating the implantation of an electronic pacemaker [39]. To create an alternative to electronic pacemakers, we sought to render electrically quiescent myocardium spontaneously active.

Our strategy to effect such a conversion was based upon the premise that ventricular myocardium contains all it requires to pace, but that pacing is normally suppressed by an expressed gene. The reasoning is as follows. In the early embryonic heart, each cell possesses intrinsic pacemaker activity. The mechanism of spontaneous beating in the early embryo is remarkably simple [78]. The opening of L-type calcium channels produces depolarization; the subsequent voltage-dependent opening of transient outward potassium channels leads to repolarization. With further development, the heart differentiates into specialized functional regions, each with its own distinctive electrical signature. The atria and ventricles become electrically quiescent; only a small number of pacemaker cells, within compact ''nodes'', set the overall rate and rhythm. Nevertheless, there is reason to wonder whether pacemaker activity may be latent within adult ventricular myocytes and masked by the differential expression of many other ionic currents. Among these, the inward rectifier potassium current (I_{K1}) is notable for its intense expression in electrically quiescent atria and ventricle, but not in nodal pacemaker cells. I_{K1}, encoded by the Kir2 gene family [38], stabilizes a strongly negative resting potential and thereby would be expected to suppress excitability. We thus explored the possibility that dominant-negative suppression of Kir2-encoded inward rectifier potassium channels in the ventricle would suffice to produce spontaneous, rhythmic electrical activity.

Replacement of three critical residues in the pore region of Kir2.1 by alanines (GYG144-146 AAA, or Kir2.1AAA) creates a dominant-negative construct [26]. The GYG motif plays a key role in ion selectivity and pore function [67]. Kir2.1AAA and GFP were packaged into a bicistronic adenoviral vector (AdEGI-Kir2.1AAA) and injected into the left ventricular cavity of guinea pigs during transient cross-clamp of the great vessels [48]. This method of delivery sufficed to achieve transduction of ~20% of ventricular myocytes. Myocytes isolated 3–4 days after in vivo transduction with Kir2.1AAA exhibited ~80% suppression of I_{K1}, but the L-type calcium current was unaffected.

Non-transduced (non-green) left ventricular myocytes isolated from AdEGI-Kir2.1AAA-injected animals, as well as green cells from AdEGI-injected hearts, exhibited no spontaneous activity, but fired single action potentials in response to depolarizing external stimuli (Fig. 1a). In contrast, Kir2.1AAA myocytes exhibited either of two phenotypes: a stable resting potential from which prolonged action potentials could be elicited by external stimuli (7 of 22 cells, not shown) or spontaneous activity (Fig. 1b). The spontaneous activity, which was seen in all cells in which I_{K1} was suppressed below 0.4 pA/pF (at –50 mV; cf. >1.5 pA/pF in controls, or 0.4–1.5 pA/pF in non-pacing Kir2.1AAA cells), resembles that of genuine pacemaker cells; the maximum diastolic potential (– 60.7 ± 2.1 mV, n = 15 of 22 Kir2.1AAA cells, $P < 0.05$ t test) is relatively depolarized, with repetitive, regular and incessant electrical activity initiated by gradual ''phase 4'' depolarization and a slow upstroke [9, 32]. Kir2.1AAA pacemaker cells responded to β-adrenergic stimulation (isoproterenol) just as SA nodal cells do, increasing their pacing rate [10, 32].

Electrocardiography revealed two phenotypes in vivo. What we most often observed was simple prolongation of the QT interval (not shown). Nevertheless, 40% of the animals exhibited an altered cardiac rhythm indicative of spontaneous ventricular foci. In normal sinus rhythm, every P wave is succeeded by a QRS complex (Fig. 1c). In two of five animals after transduction with Kir2.1AAA, premature beats of ventricular origin can be distinguished by their broad amplitude, and can be seen to ''march through'' to a beat independent of, and more rapid than, that of the physiological sinus pacemaker (Fig. 1d). In these proof-of-concept experiments, the punctate transduction required for pacing occurred by chance rather than by design, in that the distribution of the transgene throughout the ventricles was not controlled. Nevertheless, ectopic beats, arising from foci of induced pacemakers, cause the entire heart to be paced from the ventricle.

Our findings provide new insights into the biological basis of pacemaker activity. The conventional wisdom postulates that pacemaker activity requires the highly localized expression in nodal cells of ''pacemaker genes'', such as those of the HCN family [64], although an importance of scarce I_{K1} density has also been recognized [32]. Exposure to barium induces automaticity in ventricular muscle and myocytes because of its time-and voltage-dependent block of I_{K1} [27, 31]. However, barium also permeates L-type calcium channels in mixed solutions of Ca^{2+} (4 mM) and Ba^{2+} (1 mM) [62] and slows their inactivation [12], effects which make it difficult to interpret barium effects strictly in terms of I_{K1}. Our dominant-negative approach is durable and regionally specific; the barium effect is not.

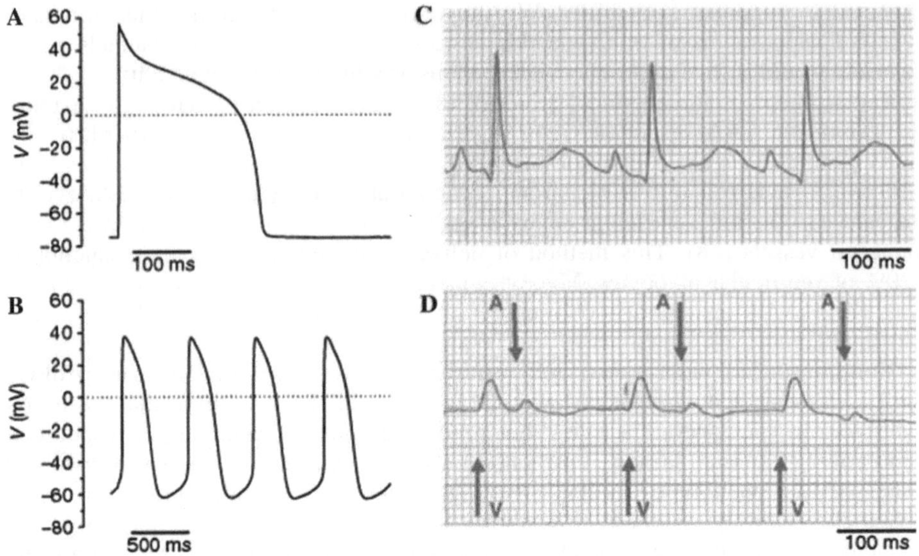

Fig. 1. Suppression of Kir2.1 channels unmasks latent pacemaker activity in ventricular cells. a Action potentials evoked by depolarizing external stimuli in control ventricular myocytes. b Spontaneous action potentials in Kir2.1AAA-transduced myocytes with depressed I_{K1}. c Baseline electrocardiograms in normal sinus rhythm. d Ventricular rhythms 72 h after gene transfer of Kir2.1AAA. P waves (*A* and *arrow*) and wide QRS complexes (*V* and *arrow*) march through to their own rhythm.

Thus, the specific suppression of Kir2 channels suffices to unleash pacemaker activity in ventricular myocytes. The crucial factor for pacing is the absence of the strongly polarizing I_{K1}, rather than the presence of special genes (although such genes may play an important modulatory role in genuine pacemaker cells) [11]. In addition to the conceptual insight into the genesis of pacing, our work implies that localized delivery of constructs such as Kir2.1AAA to the myocardium may be useful in the creation of biological pacemakers for therapeutic purposes. Focal injection into a focal area of the ventricle, possibly via an endocardial injection catheter, would be a logical means of trying to reduce this concept to practice in a larger animal.

3 Biological Pacemaker Derived from Human Embryonic Stem Cells

Human embryonic stem cells (hESCs) are pluripotent, clonogenic, and self-renewing [28]. Their versatility makes them to be one of the most effective supplies for cell-based therapies. Previous studies have demonstrated that spontaneously beating aggregates of myocytes, called embryoid bodies (EBs), could be generated from hESCs [24, 35, 50, 79]. Although these spontaneously beating human EBs (hEBs) can be derived from hESCs in vitro, they need to integrate with a recipient tissue in syncytium in order to serve as a biological pacemaker. Thus, we set out to test if the spontaneously contracting hEBs could integrate with a host tissue and thus be used as biological pacemakers [81]. First, the hESCs were stably transduced with a lentiviral construct expressing GFP as a reporter in order to locate them apart from the recipient cells by fluorescence. An in vitro transplantation model was developed in which single hESC-derived, spontaneously beating hEB (about 500 μm in diameter) was transplanted on top of a quiescent monolayer of neonatal rat ventricular myocytes (NRVMs) serving as the recipient. After 2–3 days of co-culture, synchronous rhythmic contractions of the GFP-expressing hEB and NRVM monolayer were observed at a rate of 49 ± 4 bpm (n = 14), which was similar to that of a spontaneously contracting hEB cultured alone (Fig. 2a). Observing that the transplanted hEB spontaneously contracted with the co-cultured NRVMs, we sought to examine the origin of conduction. Extracellular field potential recordings by multi-electrode array (MEA) located the site of pacemaker activity: rhythmic extracellular depolarizations were initiated from a region corresponding to the hEB-transplantation site and spread to the rest of the NRVM monolayer. Furthermore, high-resolution optical mapping further displayed a consistent time delay in action potentials recorded from the hEB to a region of NRVMs away from the transplantation site (Fig. 2b). Collectively, these observations demonstrated that the transplanted hEBs functioned as biological pacemakers driving the contraction of the recipient cells.

Since the co-culture system involved direct physical contact between the transplanted hEBs and the host NRVMs, it is possible that the rhythmic contraction of the whole co-culture was due to secondary effects rather than electrical conduction of pacemaker activity from the hEBs. We examined if effectors such as electric field potential changes [70], paracrine effects, or mechanical coupling were responsible for the electrical activities in the NRVM. First, when hEBs on a permeable plastic membrane were co-cultured with NRVMs without any physical contact, rhythmic contractions were observed only in hEBs and not in the NRVM monolayer. This excluded paracrine effects or long-range field potential change as a possible inducer of spontaneous contractility in the co-cultured NRVMs. Second, the contractility observed in the co-cultured NRVMs was not due to mere mechanical movement transduced from the spontaneously contracting hEB; spontaneous Ca^{2+} transients, not seen in otherwise-quiescent NRVMs, could be recorded from co-cultured NRVMs >1 cm away from the hEB (Fig. 2c). Use of 2,3-butanedione monoxime (BDM) is known to uncouple the excitation-contraction coupling in a myocyte [77]. Simultaneous contraction of the co-culture stopped altogether with an application of 1 mM BDM, but the electrical conduction persisted, eliminating mechanical coupling as a possible mechanism of electrical conduction from the hEBs to NRVMs.

How are, then, the spontaneous electrical depolarizations relayed to the neighboring NRVMs? Gap-junction proteins are the molecular bridges for electrical communication between cardiac cells [4, 42]. In order for the co-culture to contract spontaneously, the oscillating action potentials from the pacemaker (hEB) need to be communicated to the NRVMs by gap junction proteins. Immunostaining the co-culture with a primary antibody against the gap junction protein, connexin43 showed expression of connexin43 throughout hEB and NRVMs and along their contact surface, demonstrating the presence of gap-junctional coupling between the two tissue types. Furthermore, an application of 0.4 mM heptanol, a blocker of gap-junction proteins [33], eliminated the spontaneous contranctions in the NRVMs co-cultured

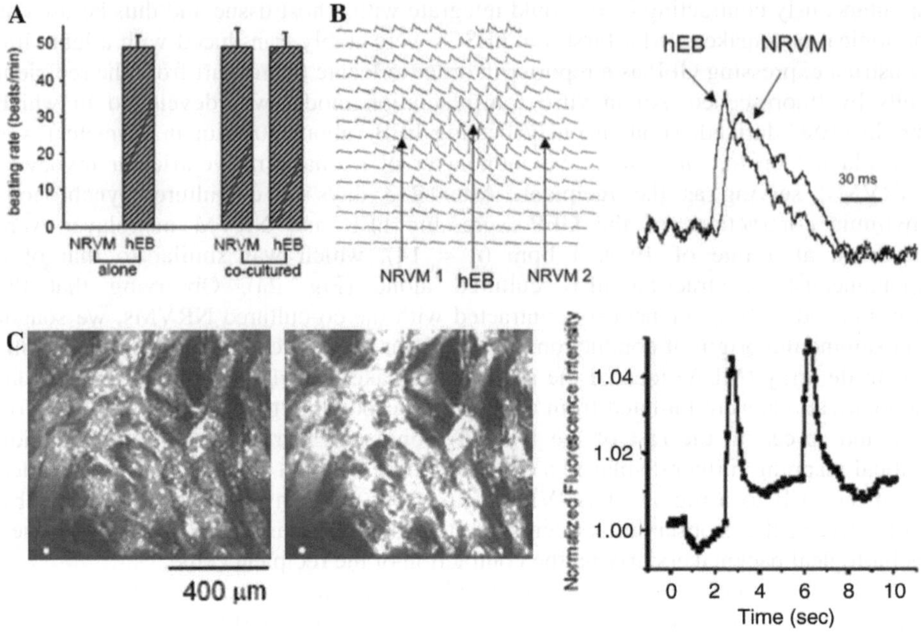

Fig.2. **a** Spontaneously beating hEB outgrowth, which stably expresses GFP, was microdissected and transplanted onto quiescent monolayer of NRVMs. Beating rate of spontaneously contracting hEB exhibited similar beating rates before (alone; 47 ± 5 bpm) and after (co-cultured; 49 ± 4 bpm) transplantation onto monolayer of NRVMs. **b** Optical APs were mapped with voltage-sensitive dye by photodiode array focused on region containing spontaneously beating hEB transplanted on quiescent NRVM monolayer (*left*). NRVM1 and NRVM2 represent two distinct sites at 3.2 and 3.6 mm, respectively, away from pacing origin. Superimposed optical AP profiles demonstrate delay of activation and slower rate of depolarization of NRVMs (*right*). **c** Ca^{2+} transient recording from NRVMs located 1 cm away from transplanted, beating hEB, with rhod-2AM as indicator before (*left*) and during (*middle*) spontaneous contraction. Normalized fluorescence intensity was measured over 10 s in co-culture (*right*).

with hEBs (Fig. 3a). Taken together, the data indicate that the spread of pacemaker activity from the hEBs to the recipient tissue proceeds via the gap-junctional coupling.

β-Adrenergic stimulation is a potent physiological mechanism to accelerate cardiac pacing [40]. We asked if the rhythmic contractions in the syncytium formed between hEBs and NRVMs could adapt its beating rates in response to a β-adrenergic agonist, isoproterenol. Indeed, the beating frequencies of the co-culture increased significantly, from 48 ± 5 to 63 ± 8 bpm, after washing in 1 μM isoproterenol ($P < 0.05$), consistent with a previous finding that β-adrenergic receptors are already expressed in hESC-derived cardiomyocytes [35]. On the other hand, bradycardiac agent ZD7288 is a specific blocker of the pacemaker ion channels, HCN [61], and would slow down the beating rate of the co-culture if the hEBs expressed the pacemaker ion channels. Indeed, addition of 100 μM ZD7288 significantly reduced the beating rate of the co-culture by fivefold (Fig. 3b; P < 0.05). Addition of neither isoproterenol nor ZD7288 affected quiescent NRVMs without engrafted hEBs. The synchronous beating could be terminated by crushing or surgically excising the transplanted human cells (n = 17; analogous to ablation), further proving that hEB pacemakers were indeed the origin of pacing.

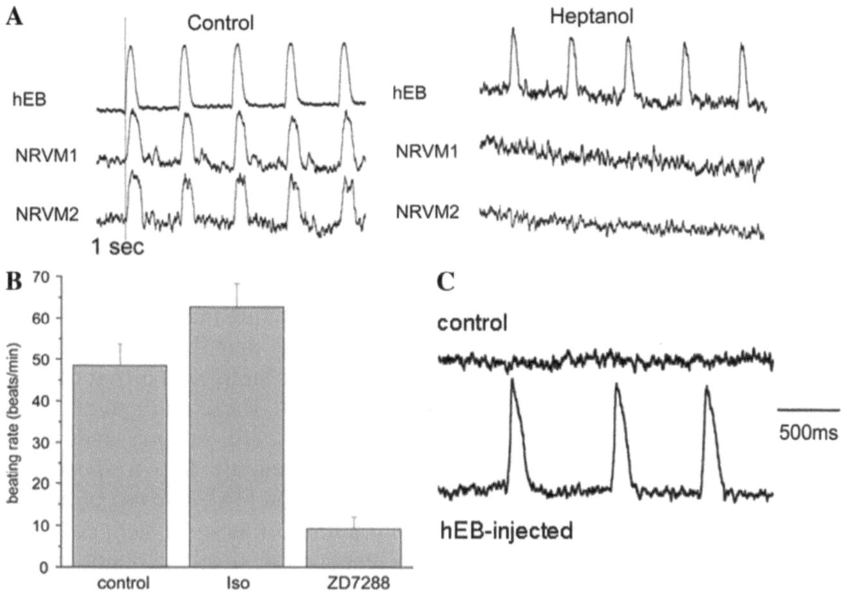

Fig. 3. **a** Gap-junction un-coupler heptanol reversibly eliminated action potential propagation to neighboring NRVM sites but did not affect APs in pacing origin of hEB at heptanol concentration of 0.4 mM. **b** β-Adrenergic stimulation with 1 μM isoproterenol (Iso) significantly accelerated spontaneous beating rate of hEB (P = 0.01), whereas ZD7288 significantly attenuated the beating activity (n = 9 for each group). **c** Optical mapping of guinea pig left ventricle pre-injected with hEBs (*bottom*) or saline (*top*) after atrioventricular nodal cryoablation without external pacing.

We then went on to examine the biological pace-maker activity in a whole heart by injecting spontaneously beating hEBs into the left ventricular anterior wall of a guinea pig in vivo. In order to distinguish ectopic ventricular beats originated from the site of injection from the animal's own sinus rhythm, the animal's endogenous SA nodal pacemaker activity was terminated by cryoablation of the atrioventricular node. Upon the ablation, ex vivo optical mapping of control (un-injected or saline-injected) guinea pig hearts exhibited complete electrical silence throughout the entire left ventricle (n = 6). However, spontaneous action potentials could be readily recorded from the epicardial surface of left ventricle of animals that had been transplanted with spontaneously beating hEBs in vivo (n = 4, Fig. 3c). Furthermore, the spontaneous action potentials were initiated from the injection site; the origin of the spread of action potentials in the epicardium coincided with the injection site of the spontaneously beating hEBs, as identified by their GFP fluorescence.

Taken together, this study demonstrates that bio-logical pacemakers derived from hESCs are capable of pacing recipient ventricular cardiomyocytes in vitro and myocardium in vivo.

4 Adult Stem Cell-Derived Biological Pacemaker

As an alternative cell source, we used adult cardiac stem cells in order to derive biological pacemakers. The heart had long been thought to be a terminally differentiated organ incapable of regeneration. The view held that the cardiomyocytes that we are born with during embryonic and fetal development do not grow in numbers but only in size. Only recently, this dogma has been challenged and refuted to form a new paradigm by the discovery of cardiac stem cells (CSCs) [5, 45–47, 54, 56]. The heart is now regarded as a self-renewing organ in which myocyte regeneration occurs throughout the organism's lifespan [2].

We have established a straightforward isolation technique that allow us to retrieve and amplify >10^6 human adult cardiac stem cells in less than 4 weeks from a single endomyocardial biopsy specimen [68]. The adult cardiac stem cells differentiated into car-diomyocytes with cardiac specific markers [68, 69]. These adult stem cells self-aggregate to form three-dimensional structures named cardiospheres and, upon co-culturing with rat ventricular myocytes, could differentiate into a spontaneously contracting cardiac tissue with innate pacemaker function [16]. The use of adult stem cells circumvents complications associated with human embryonic stem cells such as obvious ethical concerns [60], immunogenic reactions against the donor cells [21], and a visible degree of teratoma formation [53]. The autologous cell therapy using adult cardiac stem cells thus presents a unique possibility in developing biological pacemakers.

5 Creation of a Biological Pacemaker by Cell Fusion

In a previous study, human mesenchymal stem cells (hMSCs) transfected with a mouse pacemaker ion channel gene, mHCN2, were shown to induce spontaneous pacing when injected into canine left ventricular wall [57]. A key prerequisite to this approach is a high degree of gap-junctional coupling between the donor (hMSCs) and the host tissue. However, such gap junctional coupling may or may not be stable over time. Indeed, many of the major forms of human heart disease with increased arrhythmic risk coincide with gap junction remodeling and decreased cell–cell coupling [73]. In addition, frequency tuning of the stem cell-derived biological pacemaker would require further genetic manipulations. Thus, we explored the feasibility of converting normally quiescent ventricular myocytes into pacemakers by somatic cell fusion [14]. The idea is to create chemically induced fusion between myocytes and syngeneic fibroblasts engineered to express pacemaker ion channels, HCN1.

First, we established a guineapig lung fibroblast cell line stably expressing HCN1 channels with a GFP reporter (HCN1-fibroblasts). These cells were fused with freshly isolated guineapig ventricular myocytes using polyethylene glycol 1500 (PEG). The PEG-induced membrane fusion events have served as a model system to create mouse and human hybridomas [65], to study eukaryotic cell–cell fusion events [41], and to deliver outward K^+ currents into myocytes [30]. In our experiments, PEG-induced fusion occurred almost instantaneously in vitro since, within 3 min, the HCN1-fibroblasts fused with ventricular myocytes as verified by the sudden introduction of Calcein-AM fluorescence into the myocytes. The in vivo study was carried out by a simple intracardiac, focal-injection of HCN1-fibroblasts suspended in 50% PEG into the apex of guinea-pig hearts. Langendorff-isolation of ventricular myocytes from the site of injection revealed GFP-positive myocytes. We also verified in vivo fusion events by histology. HCN1-fibroblasts were first transduced with adenovirus expressing cytoplasmic β-galactosidase (Ad-*lacZ*). Immunohistochemistry against β-galactosidase and myosin heavy chain (MHC) co-localized the two proteins in regions of the myocardium, indicating fusion of cytoplasm from HCN1-fibroblasts (expressing b β-galactosidase) and cardiomyocytes.

These heterokaryons formed by in vivo fusion of myocytes and HCN1-fibroblasts verified pacemaker function by displaying spontaneous action potentials with a slow phase-4 depolarization. Biological pacemaker activities in vivo were also confirmed by electrocardiography in guinea pigs injected with HCN1-fibroblasts in PEG. Electrocardiograms recorded 1–22 days after the HCN1-fibroblast injection revealed ectopic ventricular beats that were identical in polarity and similar in morphology to those recorded during bipolar pace-mapping of the apex in the same animal (n = 5 of 13). Occasionally, junctional escape rhythms could be overtaken by ectopic ventricular pacemaker activity. Such ectopic beats were not observed in animals injected with control fibroblasts expressing GFP only (n =9).

An assumption of this study was that the fusion-induced generation of pacemaker activity is independent of cell–cell coupling. Gap-junctional coupling between cardiac fibroblasts and cardiomyocytes has been observed [37] and could provide an alternative mechanism of pacemaker activity. To test the presence of cell-coupling in our model, we loaded HCN1-fibroblasts with the membrane-impermeable dye, Calcein-AM, and mixed them with non-loaded myocytes. The dye did not diffuse from a loaded HCN1-fibroblast to the neighboring myocytes, indicating the absence of cell–cell coupling. Thus, the data indicates that the I_f-mediated pacemaker activity arises from fused heterokaryons rather than electrotonic coupling between myocytes and fibroblasts.

Comparable to the hMSC approach [57], the syngeneic fibroblasts in our study acted as a vehicle to deliver the pacemaker currents into ventricular cardiomyocytes. However, our approach is independent of gap-junctional coupling between cells and thus should be more stable in long term. Previous studies suggest that the in vivo fusion-induced heterokaryons can maintain the nuclei from each fusion partner separately and stably for at least several months [1, 22, 23, 76]. Our fusion-induced biological pacemakers were stable for at least 3 weeks and functional in less than 1 day post-injection as revealed by the electrocardiography. Furthermore, straight injection of hMSCs [57] or hESCs [36] into heart does not guarantee that the injected cells will remain at the site of injection. The fusion approach implants the biological pacemakers to the site of injection by cell fusion to cardiomyocytes. Therefore, the cell fusion approach allows us to create a biological pacing at a specific site by design rather than by chance.

6 Gene Transfer of a Synthetic Pacemaker Channel into the Heart

HCN family of channel genes figure prominently in physiological automaticity [18], and transfer of such genes into quiescent heart is the most obvious way of creating a biological pacemaker. However, use of HCN genes may be confounded by unpredictable consequences of hetero-multimerization with multiple endogenous HCN family members in the target cell [8,72]. As I_f is expressed in ventricular myocytes and can contribute to arrhythmogenesis [13, 29], HCN gene transfer in vivo may have unpredicted consequences. Similarly, little flexibility with regard to frequency tuning would be achieved if the engineered pacemaker channel co-assemble with wild-type channels upon transduction. A synthetic pacemaker channel (SPC) with no affinity to co-assemble with HCN channels would circumvent these limitations inherent with HCN gene transfer. To this end, we exploited accumulated knowledge in the biophysical properties of Shaker K^+channels. First, depolarization-activated Shaker K^+ channels had been shown to convert to a hyperpolarization-activated inward rectifier by mutating three amino acid residues in the voltage sensor (S4) of the channel [49]. Furthermore, amino acid residues in the selectivity filter of Shaker K^+ channels were found to lose its specificity and conduct Na^+ as well as K^+ when mutated to certain residues [25]. We combined the lessons from the two studies on Shaker K^+ channels and applied them to the human homologue, Kv1.4 channels [34]. By targeted mutagenesis involving <1% of the protein sequence, we were able to convert the depolarization-activated, potassium-selective human Kv1.4 channel into a hyperpolarization-activated, nonselective cation channel suitable for biological pacing applications. These mutations are comprised of three point mutations (R447N, L448A, and R453I) in the S4 segment and a single mutation (G528S) in the signature sequences of the pore's selectivity filter (Fig. 4a). The SPC channel activation by hyperpolarization and permeation by both K^+ and Na^+ ions were remarkably similar to the gating and selectivity properties of HCN channels (Fig. 4b).

An absence of heteromultimeric interactions with endogenous HCN-channels is a prerequisite to the use of such SPCs. As for the wild-type, Kv1.4 channels have previously been reported not to multimerize with the HCN channels [80]. In addition, when co-transfected into HEK293 cells, SPC did not multimerize with HCN1 as assayed by reversal potential measurements.

Next, we sought to examine in vivo pacemaker capability of SPC by creating bicistronic adenovirus expressing SPC and GFP as a reporter (AdSPC) and injected it epicardially into a guinea-pig heart. Whole-cell voltage-clamp recordings of isolated myocytes transduced with AdSPC revealed robust hyperpolarization-activated, inward currents in the presence of 0.5 mM $BaCl_2$ to eliminate I_{K1} (Fig. 4c left panel). The myocytes transduced with control adenovirus (GFP-alone) produced little inward current under identical conditions. More importantly, action potential recordings clearly demonstrated spontaneous action potentials from the AdSPC-transduced myocytes (Fig. 4c right panel, n = 7 out of 14), but not from the myocytes transduced with control adenovirus (n = 13). Furthermore, electrocardiography (ECG) taken from animals injected with AdSPC proved the in vivo pacemaker function of SPC. Monomorphic idioventricular beats could be detected in animals 3–5 days after an intracardiac injection with AdSPC (Fig. 5), but not in control animals injected with adenovirus expressing GFP alone. Collectively, these data demonstrate the feasibility of using an SPC to generate pacemaker activity in vitro and in vivo.

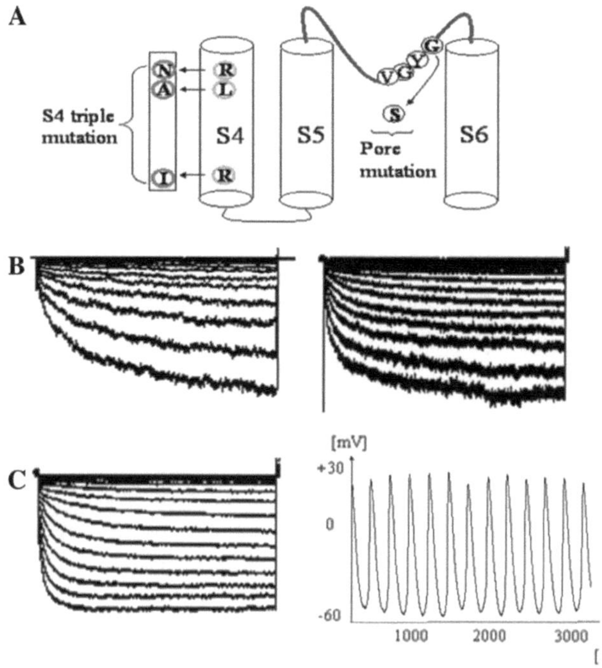

Fig. 4. a Design of synthetic pacemaker channel. To convert a human Kv1.4 channel into a pacemaker channel, three mutations (R447N, L448A, and R453I) in the S4 voltage-sensor and one pore mutation (G528S) were combined to render the channel activate upon depolarization and permeate both Na^+ and K^+, respectively. **b** Representative raw current traces expressed from HEK cells expressing SPC (left) or wild-type HCN1 channels (*right*). **c** Current trace of an AdSPC-transduced myocyte in normal Tyrode's solution (external) with 0.5 mM $BaCl_2$ (*left panel*). Spontaneous action potential (*AP*) oscillation following a triggered AP with a brief depolarizing current pulse in AdSPC-transduced myocyte. No such AP oscillations were detected in control (GFP-alone) myocytes.

Given the sparse expression of Kv1 family channels in the human heart [75] and the capability of tuning the frequency of oscillation to any given desired rate range, the synthetic pacemaker channel based on Kv1 family has potentials to be a novel therapeutic tool for use in biological pacemakers.

7 Conversion of Non-excitable Cells to Self-contained Biological Pacemakers

Most gene-or cell-based approaches in creating a biological pacemaker center around the idea of inducing spontaneous pacemaker activity in excitable but quiescent myocytes by adding pacemaker currents. An alternative is to create a pacemaker activity from non-excitable cells. We hypothesized that a non-excitable cell could be converted into a self-contained pacemaker by heterologous expression of a minimal complement of specific ion channels [15]. To this end, HEK293 cells were engineered to express the following ionic currents: (1) an excitatory current (2) an early repolarizing current, and (3) an inward rectifier current. For the excitatory current, Na^+ channel from bacteria (NaChBac) [58] was chosen because of its slow gating kinetics and its compact cDNA. A repolarizing current countering the depolarizing effects of NaChBac was provided endogenously by the HEK293 cells. Repolarizing

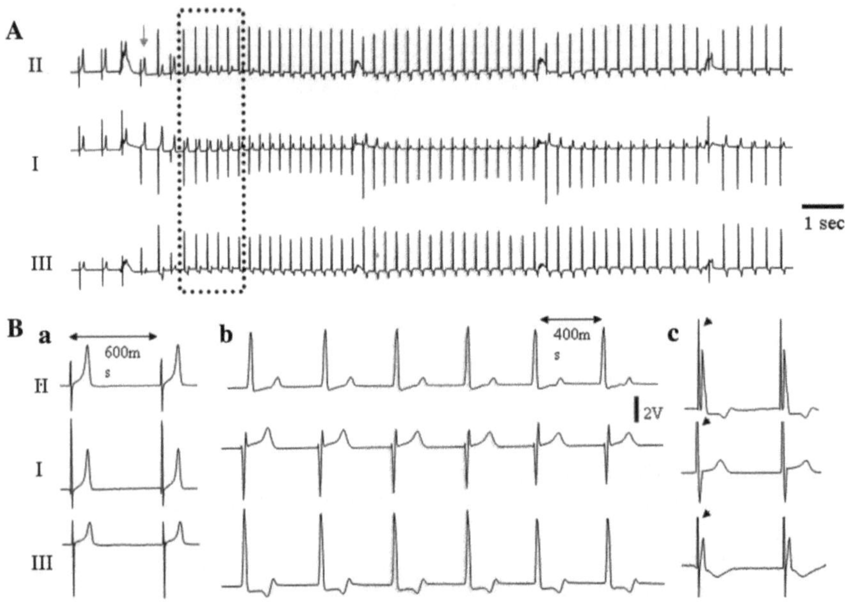

Fig.5. a An electrocardiogram showing idioventricular rhythm after the introduction of AdSPC. The *arrow* indicates start of idioventricular rhythm (150 bpm). **b** a Junctional beats are shown after the guinea pig's heart intrinsic heart rate was slowed with methacholine. b A blown-up image of idioventricular rhythm indicated as a *dashed-line square* in a. **c** Pace mapping of left ventricular free wall with an hand-held electrode. *Arrows* indicate artifacts of pacing (150 bpm). Note that idioventricular rhythms were identical in polarity to paced beats, suggesting the idioventricular rhythms were originated from left ventricular free wall.

currents were provided by a heterologous expression of Kir2.1 [38] favoring a negative diastolic potential. With only two channel genes, NaChBac and Kir2.1, expressed in HEK293 cells, action potentials could be generated in response to depolarizing external stimuli. The maximum diastolic potential (MDP) was –78 ± 7 mV with an AP duration at 90% repolarization (APD_{90}) value of 575 ± 33 ms ($n = 5$). In a previous study, a mathematical modeling based on the Luo-Rudy guinea-pig formulation suggested that addition of I_f alone could trigger a ventricular myocyte to beat spontaneously [3]. Hence, we further co-expressed HCN1 providing I_f with NaChBac and Kir2.1. Whole-cell recordings from the triple-transfected HEK cells revealed spontaneous action potentials resembling the action potential morphology of ventricular myocytes but with slow phase-4 depolarizations, a hallmark of native cardiac pacemaker cells. The spontaneous action potentials exhibited an MDP of –81.5 ± 11.8 mV, maximum rate of rise (dV/dt_{max}) of 21.6 ± 8.6 V/s, APD_{90} of 660 ± 189 ms, and a rather slow frequency of 3 ± 1 bpm which was mainly due to the slow gating kinetics of NaChBac channels (n = 4). We further combined the three channel genes, HCN1, NaChBac, and Kir2.1-GFP in tandem via internal ribosome entry site (IRES) as a polycistronic vector in a single plasmid. Expectedly, current-clamp recordings of some of the triple-gene-transfected HEK293 cells exhibited spontaneously oscillating action potentials. Therefore, this study offers a proof-of-principle that an essential set of ion channels can create spontaneous pacing activity even in non-excitable mammalian cells.

8 Future Directions

The easiest targets for near-term development are arrhythmias in which local modifications of electrical properties alone can effect treatment. Here, the singular advantages of gene/cell therapy include:

- Highly localized gene/cell delivery suffices to treat the problem. The amount of gene delivered can be correspondingly reduced, and potential problems due to widespread dissemination can be more readily averted.
- Treated cells can remain responsive to endogenous nerves and hormones. Such was the case with the biological pacemakers created by human embryonic stem cells: the biological pacemakers appropriately boosted their firing rate in response to beta-adrenergic stimulation.
- Implantable hardware is avoided, obviating long-term risks and decreasing the expense and morbidity associated with battery and lead replacements.
- The proximity to the inner lining of the heart, the endocardium, allows access by intracardiac injection, providing a potential alternative delivery route.
- The therapeutic effects can be readily detected by physical examination or by electrocardiography.
- Gene/cell transfer-induced changes can be rescued by conventional electrophysiological methods (focal ablation and pacemaker implantation).

The concepts are generalizable to ventricular arrhythmias such as those associated with heart failure or heritable long QT syndrome. In heart failure, for example, over-expression of K channels can be used to antagonize the acquired long QT syndrome [51, 52]; the attendant loss of contractility may be amenable to co-administration of a second gene to augment calcium cycling, in a dual gene therapy strategy. While such work is conceptually attractive, widespread delivery with long-term expression will be required before human trials can be anticipated.

Sizeable work is ahead in terms of fine-tuning the frequency, toxicology of the gene and cells used, malignancy of the stem cells, and long-term stability of biological pacemakers. Nevertheless, given the promises, the effort to develop biological alternatives to the present therapies is justified.

References

1. Alvarez-Dolado, Pardal R, Garcia-Verdugo JM et al (2003) Fusion of bone-marrow-derived cells with Purkinje neurons, cardiomyocytes and hepatocytes. Nature 425(6961):968–973
2. Anversa P, Kajstura J, Leri A, Bolli R (2006) Life, death of cardiac stem cells: a paradigm shift in cardiac biology. Circulation 113(11):1451–1453
3. Azene EM, Xue T, Marban E, Tomaselli GF, Li RA (2005) Non-equilibrium behavior of HCN channels: insights into the role of HCN channels in native and engineered pacemakers. Cardiovasc Res 67(2):263–273
4. Barr L, Dewey MM, Berger W (1965) Propagation of action potentials and the structure of the nexus in cardiac muscle. J Gen Physiol 48:797–823
5. Beltrami AP, Barlucchi L, Torella D et al (2003) Adult cardiac stem cells are multipotent and support myocardial regeneration. Cell 114(6):763–776
6. Bernstein AD, Parsonnet V (2001) Survey of cardiac pacing and implanted defibrillator practice patterns in the United States in 1997. Pacing Clin Electrophysiol 24(5):842–855
7. Blaese RM, Culver KW, Miller AD et al (1995) T lymphocyte-directed gene therapy for ADA-SCID: initial trial results after 4 years. Science 270(5235):475–480
8. Brewster AL, Bernard JA, Gall CM, Baram TZ (2005) Formation of heteromeric hyperpolarization-activated cyclic nucleotide-gated (HCN) channels in the hippocampus is regulated by developmental seizures. Neurobiol Dis 19(1– 2):200–207
9. Brooks CM, Lu H-h (1972) The sinoatrial pacemaker of the heart. Charles C. Thomas, Springfield
10. Brown HF, McNaughton PA, Noble D, Noble SJ (1975) Adrenergic control of cardian pacemaker currents. Philos Trans R Soc Lond B Biol Sci 270(908):527–537
11. Brown HF, Kimura J, Noble D, Noble SJ, Taupignon A (1984) The ionic currents underlying pacemaker activity in rabbit sino-atrial node: experimental results and computer simulations. Proc R Soc Lond B Biol Sci 222(1228):329–347
12. Campbell DL, Giles WR, Shibata EF (1988) Ion transfer characteristics of the calcium current in bull-frog atrial myocytes. J Physiol 403:239–266
13. Cerbai E, Pino R, Porciatti F et al (1997) Characterization of the hyperpolarization-activated current, I(f), in ventricular myocytes from human failing heart. Circulation 95(3):568–571
14. Cho HC, Kashiwakura Y, Marban E (2007) Creation of a biological pacemaker by cell fusion. Circ Res 100(8): In Press.
15. Cho HC, Kashiwakura Y, Marban E (2005) Conversion of non-excitable cells to self-contained biological pacemakers. Circulation 112(17):II-307
16. Cho HC, Smith R, Abraham M, Messina E, Giacomello A, Marban E (2005) Electrophysiology of human and porcine adult cardiac stem cells isolated from endomyocardial biopsies. Heart Rhythm 2(5):S45
17. Coplen SE, Antman EM, Berlin JA, Hewitt P, Chalmers TC (1990) Efficacy and safety of quinidine therapy for maintenance of sinus rhythm after cardioversion. A meta-analysis of randomized control trials. Circulation 82(4):1106–1116

18. DiFrancesco D (1995) The pacemaker current (I(f)) plays an important role in regulating SA node pacemaker activity. Cardiovasc Res 30(2):307–308
19. Echt DS, Liebson PR, Mitchell LB et al (1991) Mortality and morbidity in patients receiving encainide, flecainide, or placebo. The cardiac arrhythmia suppression trial. N Engl J Med 324(12):781–788
20. Falk RH (2001) Atrial fibrillation. N Engl J Med 344(14):1067–1078
21. Gerecht-Nir S, Itskovitz-Eldor J (2004) Cell therapy using human embryonic stem cells. Transpl Immunol 12(3–4):203– 209
22. Gibson AJ, Karasinski J, Relvas J et al (1995) Dermal fibroblasts convert to a myogenic lineage in mdx mouse muscle. J Cell Sci 108(Pt 1):207–214
23. Gussoni E, Bennett RR, Muskiewicz KR et al (2002) Long-term persistence of donor nuclei in a Duchenne muscular dystrophy patient receiving bone marrow transplantation. J Clin Invest 110(6):807–814
24. He JQ, Ma Y, Lee Y, Thomson JA, Kamp TJ (2003) Human embryonic stem cells develop into multiple types of cardiac myocytes: action potential characterization. Circ Res 93(1):32–39
25. Heginbotham L, Lu Z, Abramson T, MacKinnon R (1994) Mutations in the K+ channel signature sequence. Biophys J 66(4):1061–1067
26. Herskowitz I (1987) Functional inactivation of genes by dominant negative mutations. Nature 329(6136):219–222
27. Hirano Y, Hiraoka M (1988) Barium-induced automatic activity in isolated ventricular myocytes from guinea-pig hearts. J Physiol 395:455–472
28. Hoffman LM, Carpenter MK (2005) Characterization and culture of human embryonic stem cells. Nat Biotechnol 23(6):699–708
29. Hoppe UC, Jansen E, Sudkamp M, Beuckelmann DJ (1998) Hyperpolarization-activated inward current in ventricular myocytes from normal, failing human hearts. Circulation 97(1):55–65
30. Hoppe UC, Johns DC, Marban E, O'Rourke B (1999) Manipulation of cellular excitability by cell fusion: effects of rapid introduction of transient outward K+ current on the guinea pig action potential. Circ Res 84(8):964–972
31. Imoto Y, Ehara T, Matsuura H (1987) Voltage-and time-dependent block of iK1 underlying Ba2+-induced ventricular automaticity. Am J Physiol 252(2 Pt 2):H325–H333
32. Irisawa H, Brown HF, Giles W (1993) Cardiac pacemaking in the sinoatrial node. Physiol Rev 73(1):197–227
33. Jalife J, Sicouri S, Delmar M, Michaels DC (1989) Electrical uncoupling and impulse propagation in isolated sheep Purkinje fibers. Am J Physiol 257(1 Pt 2):H179–H189
34. Kashiwakura Y, Cho HC, Azene E, Marban E (2005) Creation of a synthetic pacemaker channel. Circulation 114(16):1682–1686
35. Kehat I, Kenyagin-Karsenti D, Snir M et al (2001) Human embryonic stem cells can differentiate into myocytes with structural and functional properties of cardiomyocytes. J Clin Invest 108(3):407–414
36. Kehat I, Khimovich L, Caspi O et al (2004) Electromechanical integration of cardiomyocytes derived from human embryonic stem cells. Nat Biotechnol 22(10):1282–1289
37. Kohl P (2003) Heterogeneous cell coupling in the heart: an electrophysiological role for fibroblasts. Circ Res 93(5):381–383
38. KuboY, Baldwin TJ, Jan YN, Jan LY (1993) Primary structure and functional expression of a mouse inward rectifier potassium channel. Nature 362(6416):127–133
39. Kusumoto FM, Goldschlager N (1996) Cardiac pacing. N Engl J Med 334(2):89–97

40. Lakatta EG, Maltsev VA, Bogdanov KY, Stern MD, Vinogradova TM (2003) Cyclic variation of intracellular calcium: a critical factor for cardiac pacemaker cell dominance. Circ Res 92(3):e45–e50
41. Lentz BR, Lee JK (1999) Poly(ethylene glycol) (PEG)-mediated fusion between pure lipid bilayers: a mechanism in common with viral fusion and secretory vesicle release? Mol Membr Biol 16(4):279–296
42. Loewenstein WR (1981) Junctional intercellular communication: the cell-to-cell membrane channel. Physiol Rev 61(4):829–913
43. Losordo DW, Vale PR, Symes JF et al (1998) Gene therapy for myocardial angiogenesis: initial clinical results with direct myocardial injection of phVEGF165 as sole therapy for myocardial ischemia. Circulation 98(25):2800–2804
44. Mann MJ, Whittemore AD, Donaldson MC et al (1999) Exvivo gene therapy of human vascular bypass grafts with E2F decoy: the PREVENT single-centre, randomised, controlled trial. Lancet 354(9189):1493–1498
45. Martin CM, Meeson AP, Robertson SM et al (2004) Persistent expression of the ATP-binding cassette transporter, Abcg2, identifies cardiac SP cells in the developing and adult heart. Dev Biol 265(1):262–275
46. Matsuura K, Nagai T, Nishigaki N et al (2004) Adult cardiac Sca1-positive cells differentiate into beating cardiomyocytes. J Biol Chem 279(12):11384–11391
47. Messina E, De Angelis L, Frati G et al (2004) Isolation and expansion of adult cardiac stem cells from human and murine heart. Circ Res 95(9):911–921
48. Miake J, Marban E, Nuss HB (2002) Biological pacemaker created by gene transfer. Nature 419(6903):132–133
49. Miller AG, Aldrich RW (1996) Conversion ofa delayed rectifier K+ channel to a voltage-gated inward rectifier K+ channel by three amino acid substitutions. Neuron 16(4):853–858
50. Mummery C, Ward-van Oostwaard D, Doevendans P et al (2003) Differentiation of human embryonic stem cells to cardiomyocytes: role of coculture with visceral endoderm-like cells. Circulation 107(21):2733–2740
51. Nuss HB, Johns DC, Kaab S et al (1996) Reversal of potassium channel deficiency in cells from failing hearts by adenoviral gene transfer: a prototype for gene therapy for disorders of cardiac excitability and contractility. Gene Ther 3(10):900–912
52. Nuss HB, Marban E, Johns DC (1999) Overexpression of a human potassium channel suppresses cardiac hyperexcit-ability in rabbit ventricular myocytes. J Clin Invest 103(6):889–896
53. Odorico JS, Kaufman DS, Thomson JA (2001) Multilineage differentiation from human embryonic stem cell lines. Stem Cells 19(3):193–204
54. Oh H, Bradfute SB, Gallardo TD et al (2003) Cardiac progenitor cells from adult myocardium: homing, differentiation, and fusion after infarction. Proc Natl Acad Sci USA 100(21):12313–12318
55. Ohno T, Gordon D, San H et al (1994) Gene therapy for vascular smooth muscle cell proliferation after arterial injury. Science 265(5173):781–784
56. Pfister O, Mouquet F, Jain M et al (2005) CD31– but Not CD31+ cardiac side population cells exhibit functional cardiomyogenic differentiation. Circ Res 97(1):52–61
57. Potapova I, Plotnikov A, Lu Z et al (2004) Human mesenchymal stem cells as a gene delivery system to create cardiac pacemakers. Circ Res 94(7):952–959
58. Ren D, Navarro B, Xu H, Yue L, Shi Q, Clapham DE (2001) A prokaryotic voltage-gated sodium channel. Science 294(5550):2372–2375

59. Richardson AW, Josephson ME (1999) Ablation of ventricular tachycardia in the setting of coronary artery disease. Curr Cardiol Rep 1(2):157–164
60. Robertson JA (2001) Human embryonic stem cell research: ethical and legal issues. Nat Rev Genet 2(1):74–78
61. Robinson B, Siegelbaum SA (2003) Hyperpolarization-activated cation currents: from molecules to physiological function. Annu Rev Physiol 65:453–480
62. Rodriguez-Contreras A, Nonner W, Yamoah EN (2002) Ca2+ transport properties and determinants of anomalous mole fraction effects of single voltage-gated Ca2+ channels in hair cells from bullfrog saccule. J Physiol 538(Pt 3):729–745
63. Rosengart TK, Lee LY, Patel SR et al (1999) Six-month assessment of a phase I trial of angiogenic gene therapy for the treatment of coronary artery disease using direct intramyocardial administration of an adenovirus vector expressing the VEGF121 cDNA. Ann Surg 230(4):466–470; discussion 70–72
64. Santoro B, Tibbs GR (1999) The HCN gene family: molecular basis of the hyperpolarizationactivated pacemaker channels. Ann N Y Acad Sci 868:741–764
65. Shirahata S, Katakura Y, Teruya K (1998) Cell hybridization, hybridomas, and human hybridomas. Methods Cell Biol 57:111–145
66. Siebels J, Cappato R, Ruppel R, Schneider MA, Kuck KH (1993) Preliminary results of the Cardiac Arrest Study Hamburg (CASH). CASH Investigators. Am J Cardiol 72(16):109F–113F
67. Slesinger PA, Patil N, Liao YJ, Jan YN, Jan LY, Cox DR (1996) Functional effects of the mouse weaver mutation on G proteingated inwardly rectifying K^+ channels. Neuron 16(2):321–331
68. Smith R, Messina E, Abraham M, Cho HC, Giacomello A, Marban E (2005) Electrophysiology of human and porcine adult cardiac stem cells isolated from endomyocardial biopsies. Circulation 111(13):1720
69. Smith R, Barile L, Cho HC et al (2005) Unique phenotype of cardiospheres derived from human endomyocardial biopsies. Circulation 112(17):II-51
70. Sperelakis N (2002) An electric field mechanism for transmission of excitation between myocardial cells. Circ Res 91(11):985–987
71. Thom T, Haase N, Rosamond W et al (2006) Heart disease and stroke statistics-2006 update: a report from the American Heart Association Statistics Committee and Stroke Statistics Subcommittee. Circulation 113(6):e85–e151
72. Ulens C, Tytgat J (2001) Functional heteromerization of HCN1 and HCN2 pacemaker channels. J Biol Chem 276(9):6069–6072
73. van der Velden HM, Jongsma HJ (2002) Cardiac gap junctions and connexins: their role in atrial fibrillation and potential as therapeutic targets. Cardiovasc Res 54(2):270–279
74. Waldo AL, Camm AJ, deRuyter H et al (1996) Effect of d-sotalol on mortality in patients with left ventricular dys-function after recent and remote myocardial infarction. The SWORD investigators. Survival with oral d-sotalol. Lancet 348(9019):7–12
75. Wang Z, Feng J, Shi H, Pond A, Nerbonne JM, Nattel S (1999) Potential molecular basis of different physiological properties of the transient outward K+ current in rabbit and human atrial myocytes. Circ Res 84(5):551–561
76. Weimann JM, Johansson CB, Trejo A, Blau HM (2003) Stable reprogrammed heterokaryons form spontaneously in Purkinje neurons after bone marrow transplant. Nat Cell Biol 5(11):959–966
77. WierWG, Blatter LA (1991) Ca(2+)-oscillations and Ca(2+)-waves in mammalian cardiac and vascular smooth muscle cells. Cell Calcium 12(2–3):241–254

78. Wobus AM, Rohwedel J, Maltsev V, Hescheler J (1995) Development of cardiomyocytes expressing cardiac-specific genes, action potentials, and ionic channels during embryonic stem cell-derived cardiogenesis. Ann N Y Acad Sci 752:460–469
79. Xu C, Police S, Rao N, Carpenter MK (2002) Characterization and enrichment of cardiomyocytes derived from human embryonic stem cells. Circ Res 91(6):501–508
80. Xue T, Marban E, Li RA (2002) Dominant-negative suppression of HCN1-and HCN2-encoded pacemaker currents by an engineered HCN1 construct: insights into structure-function relationships and multimerization. Circ Res 90(12):1267–1273
81. Xue T, Cho HC, Akar FG et al (2005) Functional integration of electrically active cardiac derivatives from genetically engineered human embryonic stem cells with quiescent recipient ventricular cardiomyocytes: insights into the development of cell-based pacemakers. Circulation 111(1):11–20

Creating a Cardiac Pacemaker by Gene Therapy

Traian M. Anghel and Steven M. Pogwizd

Department of Medicine, Section of Cardiology,
University of Illinois at Chicago, 840 South Wood Street,
M/C 715, Chicago, IL 60612, USA
spogwizd@uic.edu

Abstract. While electronic cardiac pacing in its various modalities represents standard of care for treatment of symptomatic bradyarrhythmias and heart failure, it has limitations ranging from absent or rudimentary autonomic modulation to severe complications. This has prompted experimental studies to design and validate a biological pacemaker that could supplement or replace electronic pacemakers. Advances in cardiac gene therapy have resulted in a number of strategies focused on β-adrenergic receptors as well as specific ion currents that contribute to pacemaker function. This article reviews basic pacemaker physiology, as well as studies in which gene transfer approaches to develop a biological pacemaker have been designed and validated in vivo. Additional requirements and refinements necessary for successful biopacemaker function by gene transfer are discussed.

1 Introduction

The natural prototype of a pacemaker is the sinoatrial node (SAN). The innate hierarchy in the cardiac conduction system allows for lower centers [at different levels in the SAN, atrioventricular node (AVN), Purkinje fibers, atria, and ventricles] to take over as pacemakers should their immediately superior center fail. If this chronotropic failure, whether from conduction block or other mechanisms, achieves clinical significance in a patient, it is part of standard medical practice for physicians in economically developed countries to resort to the implantation of an electronic cardiac pacemaker [29]. Since the first implant of a pacemaker in a human by Dr. Ake Senning in 1958 [24], the device has evolved greatly in terms of miniaturization, longevity, and complexity of functions performed. Over 600,000 pacemakers are implanted annually in the world [51]. Of these, more than 200,000 are in the USA alone, where it has been estimated that in excess of one million people are reliant on permanent implantable pacemakers. Despite the numerous improvements in technology incorporated into these devices and in the implant techniques, the electronically driven system is not innocuous. Pacemaker malfunction may be related to abnormal sensing of electrical activity or failure to capture. The most feared pacemaker complications are infections, embolic events, and mechanical dysfunction, which, along with battery depletion, may require removal and replacement of part or all of the pacemaker system.

On this background, the impetus for a biological pacemaker, that could ultimately supplement and/or replace electronic pacemakers, has been burgeoning. Our understanding of pacemaker physiology has grown to a level where hypothesis-driven

J.A.E. Spaan (Ed.): Biopacemaking, BIOMED, pp. 45–62, 2007.
springerlink.com

experimentation has become possible. Advances in cardiac gene therapy [3, 20, 21] have led to a variety of strategies to produce pacemaker function in vivo, by delivering native DNA or adenoviruses encoding particular genes into myocardial cells. Here we review normal sinus node physiology, and the ion currents and modulators that contribute to normal pacemaker function. This is followed by a review of recent gene transfer strategies to create a biological pacemaker in vivo including their successes, their limitations, and their future directions.

2 Pacemaker Physiology

The SAN is a very complex, heterogeneous structure, despite its relatively small size. Its anatomical and electrical architecture have evolved such that the SAN manages to drive the much larger mass of surrounding atrium without being in turn suppressed by it. There are a number of ionic currents that underlie SAN (and AVN) pacemaker activity (see below). Moreover, the anatomic architecture involving a connective tissue barrier [69], the presence of transitional cells [13, 69] and some degree of electrical uncoupling within the SAN [1, 5, 46] also contribute to normal pacemaker physiology. The complex structure of the SAN accommodates populations of cells, different morphologically and functionally, that coexist to yield pacemaking activity, responsiveness to a large spectrum of humoral and neural influences, and protection from unwarranted influences. This would imply that the engineering of a functional biological pacemaker will most likely require the creation of a complex of genetically modified cells rather than just of a single cell line. Of course, as the final aim is not necessarily the recreation of the SAN, the final structure of a viable, functional construct would not have to reproduce in detail the complex architecture of the natural pacemaker (connective tissue, transitional cells, current gradients etc.). Also, just as in the case of the electronic pacemaker, it may be necessary to replicate or mimic just a basic set of the functions performed by the SAN (such as generation of a pacemaker current, the ability to drive the surrounding cardiomyocytes, the autonomy of pacemaking activity in the electrotonic milieu of the surrounding tissues, etc.). It remains to be seen as to whether these important characteristics of the SAN can be achieved in a manmade biological implantable pacemaker using gene transfer approaches. For now, the ionic currents that drive the pacemaker seem the most suitable targets for further experimentation and validation.

3 Normal Pacemaker Physiology and Ion Channel Candidates for a Stable Biological Pacemaker

During phase 4, cells in the SAN, AVN and the His-Purkinje system undergo slow diastolic depolarization. No single current is entirely responsible for this. Rather, this depolarization is the result of a complex interplay of various ionic currents, with marked interspecies variability noted [31, 32, 65, 79]. The currents known to date to be active, under different experimental conditions, in the spontaneously beating mammalian SAN cells include (Fig. 1): the hyperpolarization-activated inward cur rent I_f; the L-type calcium current, I_{CaL} and the closely related sustained inward current I_{st} [65]; the T-type calcium current, I_{CaT}; the sodium–calcium exchanger,$I_{Na/Ca}$; the delayed rectifier current, I_K [62] (in toto, or through its individual components: slow, I_{Ks}, fast, I_{Kr}, or ultrafast, I_{Kur}) [30, 62]; the transient outward current I_{to}; and the inward rectifying I_{K1}. Despite a considerable amount of work on the subject, by many groups, spanning many decades, it is still controversial as to which of the above currents, if any, dominates the pacemaking activity of the SAN under normal in vivo conditions, or as to whether the pacemaker activity of the SAN is indeed a solely, membranebased, phenomenon. From the vast body of knowledge adduced, to date, for the currently prevalent theory, that of a membranary pacemaker, the most likely candidates for the role of the still elusive, dominant, pacemaker current appear to be I_f and the calcium currents.

I_f the funny current, is carried by the hyperpolarization-activated, cyclic nucleotide gated (HCN) channel, which, through a cyclic AMP (cAMP) binding site, is sensitive to modulation by catecholamines. I_f is well accepted as a component of the pacemaker current present in the SAN, and at various levels in the AVN, Purkinje fibers, atria, and the ventricles [81]. The magnitude to which this current participates in the pacemaking activity of the cardiac conduction system, though, has been a subject of intense debate [6, 16–18, 66, 79]. DiFrancesco, Brown, and their colleagues [8, 14, 15] have contributed a wealth of data suggesting a cardinal role of I_f in the initiation, and the control thereafter, of the rate of diastolic depolarization. The principal observations supporting this role are: (1) expression of I_f parallels the presence of spontaneous activity of cardiomyocytes of adult mammals, (2) during patch clamping experiments, the partially depolarized, nonbeating SAN cells start beating during hyperpolarizing steps, and not during depolarization, and (3) the rate of SAN cells accelerates with application of more hyperpolarizing voltages [16] (Fig. 2). Data from other groups suggest that the role of I_f in the highly structured SAN may be very important in achieving other functions. As described by Boyett et al. [7] the increased density of I_f toward the periphery of the node, while regarded as insufficient to carry the pacemaking current by itself [79], seems to protect the SAN from the hyperpolarizing influence of the right atrium. The grounds for this theory lie on voltage clamp data. I_f is activated, under experimental conditions, by pulses achieving hyperpolarization of –70 mV or more, which is greater than the middiastolic potential of the SAN cells, but similar to that of the right atrial myocytes. Thus, at the border of the SAN, the hyperpolarizing atrial influence would induce increased activation of I_f, which in turn opposes the hyperpolarization, thus protecting the electrical milieu present in the more central regions of the SAN proper.

Since the discovery of the voltage-dependent calcium channels in the rabbit SAN, and the initial thorough description by Hagiwara's group in 1988 [33, 60], ample direct and indirect evidence has accumulated as to the contribution of these channels to the diastolic depolarization of pacemaker cells. Identified in tissues exhibiting

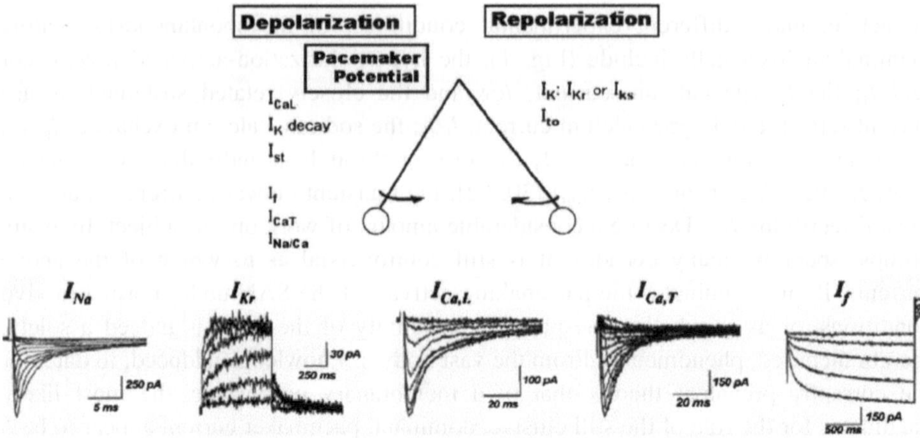

Fig. 1. *Top*: diagram of the ionic mechanisms and membrane currents involved in the generation and control of the pendulum of spontaneous electrical activity of the SAN cells. Reprinted from Ref. 79, with permission. *Bottom*: representative tracings of ionic currents recorded in spontaneously beating murine SAN cells. Reprinted from Ref. 60, with permission.

automaticity [26, 35, 36], I_{CaL} is the main pacemaking current recorded at the center of the SAN [7, 48]. Early on, it was described as a dihydropyridinesensitive [48, 80, 87] current, activated at about –30 mV and responsive to catecholamines [41]. I_{CaT}, too, was proposed as a driver of the SAN diastolic depolarization [33, 41]. Described as a high threshold current, typically sensitive to blockade by Ni^{2+}, with activation in the more hyperpolarized range (–50 mV), I_{CaT} appeared to be lacking sensitivity to both catecholamines and calcium channel blockers.

We now know that the different types of calcium channels are largely defined by their respective pore forming α_1 subunit, of which ten have been catalogued to date by means of homology screening. They are grouped in three related families (Ca_v1, Ca_v2, and Ca_v3) [25], whose function has been thoroughly characterized [38]. $Ca_v1.2$ and $Ca_v1.3$ underlie the cardiac I_{CaL} [38, 58, 61], with the latter having a particularly lower activation threshold, slower inactivation, and lower sensitivity to dihydropyridines than the former. Following findings, by Platzer et al. [70], that $Ca_v1.3$ knockout mice had significantly altered SAN function, further work by others suggested that $Ca_v1.3$ is particularly important in the genesis of the pacemaker current in the SAN and in the AVN [33, 59, 63, 91]. Also, indirect evidence is pointing to $Ca_v1.3$ as the subunit responsible for cathecholaminergic sensitivity of I_{CaL} [63] Altogether, these data establish $Ca_v1.3$, distinct from other Ca_v1 subfamilies, as the ''pacemaker'' form of I_{CaL} [60] which is responsible for the control of contraction and possibly for the upstroke phase of the action potential [59, 63].

Classically, and perhaps, at times, dogmatically, cardiac pacemaking has been described as an exclusively membrane delimited mechanism. Speaking to a higher degree of complexity of the natural pacemaker is an increasing evidence on the role the so-called ''sarcoplasmic reticulum Ca^{2+} clock'' (local calcium release, LCR, and *spontaneous calciumreleaseignited excitation*, SCaRIE) [11, 57]. A novel

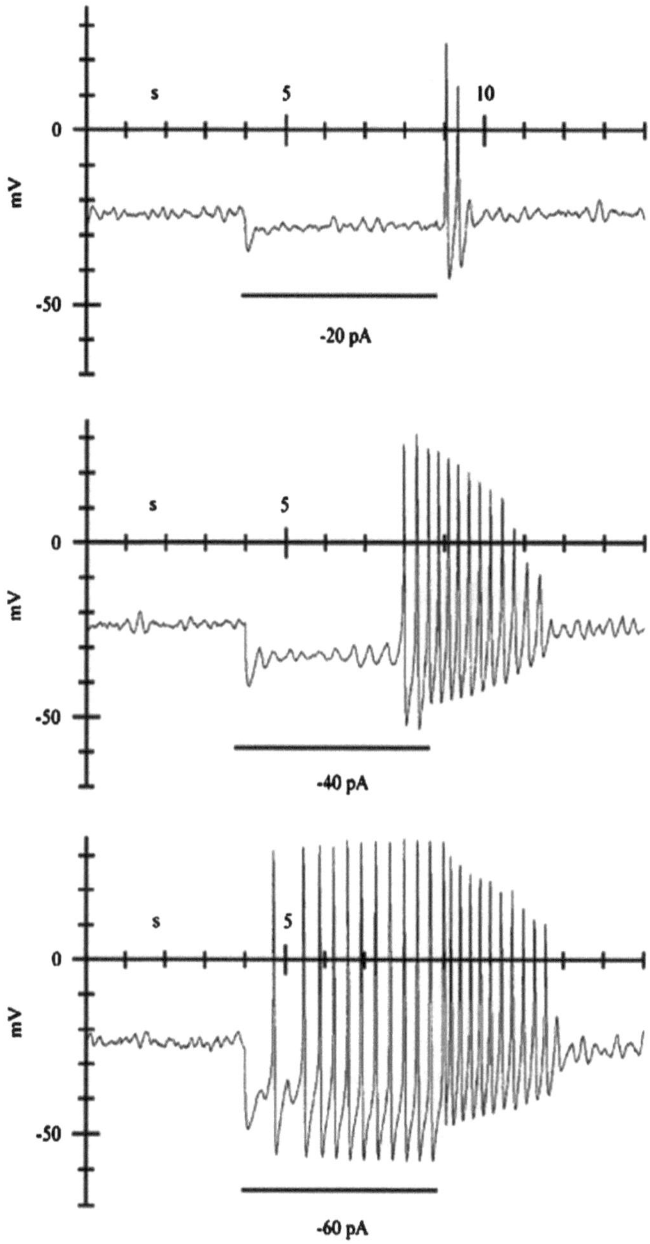

Fig. 2. Involvement of I_f in pacemaking: hyperpolarization triggers electrical activity. Application of hyperpolarizing steps to a depolarized SAN cell shows activation of I_f starting at –40 mV. Electrical activity becomes more stable and persistent at more negative voltages. Reprinted from Ref. 16, with permission.

hypothesis [57] on the functioning of the normal pacemaker has been proposed, by integrating, via activation of I_{CaL} (as the common event), the "external loop" of the "membrane clock" (I_k, I_{st}, I_f, I_{CaT}, I_{CaL}) with the "internal loop" of the intracellular calcium clock. Briefly, by this theory [57], backed up by solid experimental data [4, 52, 56, 88], the SAN cells calcium clock is both entrained by an action potential (during early diastolic depolarization) and "free running" (during the later part of the preaction potential phase, when LCRs of significant frequency and strength initiate the SAN "duty cycle" [57] by SCaRIE of the NCX, and subsequent initiation of an action potential by the I_{CaT}) [88]. A more detailed discussion of these mechanisms is presented in other sections of this journal, but it is clear that a *unified theory* of pacemaking is still in the making. The role of the calcium clock has been downplayed or ignored in most of the theoretical pacemaker models, and it is expected that its inclusion, along with its characteristic modulation by PKA-dependent phosphorylation, in such models will broaden our theoretical armamentarium in the quest for the creation of a biological pacemaker. To date, though, there have been no attempts, to our knowledge, of genetic manipulation of submembrane calcium mechanisms for the purpose of obtaining a biopacemaker per se.

4 Autonomic Modulation of Pacemaker Activity

As noted above, the main pacemaker current carriers, I_f and I_{CaL} are modulated by the sympathetic and parasympathetic nervous systems, but this neurohumoral influence on the rate of discharge of the pacemaker cells is more concerted. In principle, the discharge rate may be increased by lowering the maximum diastolic potential (MDP), increasing the slope of phase 4, or increasing the threshold potential; the opposite changes may decelerate discharge rate. Sympathetic and parasympathetic activation alters SAN pacemaker rates in two ways: (a) they shift the primary site of the pacemaker within the SAN, and (b) they modify the kinetics of the currents that control pacemaker activity. Sympathetic stimulation, mediated primarily by β-adrenergic-receptor (β-AR) activation, has a positive chronotropic effect through a shift of the site of the primary pacemaker toward the cephalad region of the SAN and by increasing I_{CaT}, I_f, I_{st} and I_{ks} [7, 16, 79]. Acetylcholine and vagal stimulation have the opposite effect on the activation of I_f and I_{CaL}, and the negative chronotropic effect is enhanced by I_{KAch} activation and caudal displacement of the primary pacemaker site within the SAN. Thus, attempts at genetically modifying the influence of the autonomic nervous system on preexistent structures of the cardiac conduction system implicitly rely on normal innate pacemaking structures. Enhancing the autonomic influence of a diseased pacemaker could conceivably have deleterious effects such as arrhythmias. It is thus likely that this approach would have a rather peripheral role in the synthesis of a biopacemaker.

5 Gene Transfer to the Heart

Recent advances in cardiac gene therapy have shown great promise for the eventual treatment of cardiac disease [3, 20, 21]. Moreover, cardiac gene transfer has been used to modulate the electrophysiologic substrate in a number of disease settings. Gene transfer of repolarizing K channel proteins has been used to modulate repolarization, and the results to date suggest that action potential duration (APD) can be significantly shortened by such approaches [42–44, 53, 67, 68]. More recently, Donahue and colleagues [2, 19, 20] performed adenoviral gene transfer of an inhibitory G protein α-subunit ($G_{\alpha i2}$) into the AV node of swine in vivo, and demonstrated successful control of the ventricular rate response in chronic atrial fibrillation. The success of these proof-of-principle studies suggest that gene transfer approaches may be useful for the development of a biological pacemaker.

6 Gene Transfer Approaches for Biopacemakers

Initial gene transfer approaches for a biologic pacemaker were aimed at altering the adrenergic input to the native cardiac pacemaker(s) by transferring adrenergic receptors to the surrounding regions [22, 23]. β-ARs regulate cardiac chronotropic (as well as inotropic) response by G-protein coupled signaling pathways—in part through modulation of I_f [37, 40]. Edelberg et al. [22] first explored the β_2-AR as a target for gene transfer, choosing the human β_2-AR because it is immunologically distinct from, but functionally and structurally similar to, the murine receptor. They demonstrated that in vivo injection of the β_2-AR construct into the right atrium of mouse hearts increased the endogenous heart rate by ~40% (vs. controlled injected hearts) by 2 days, after which the heart rate returned to baseline (by day 7). In a subsequent study [23], they showed that fluoroscopically guided injection of the β_2-AR construct in the right atrium of pigs resulted in a heart rate increase of 50% (without apparent change in atrial conduction) and expression of the encoded β_2–AR gene (by immunostaining). These short-term studies demonstrated the feasibility of modulating cardiac pacemaker activity by gene transfer approaches in mouse and large animal models, critical proof-of-principle showing the potential for gene transfer on modulating heart rate in vivo. From a practical point of view, limitations to these studies included the use of recombinant cDNA construct (rather than viral gene transfer), the short-lived effects, and non-specific yield (as the gene transfer affected not only the activity of the pacemaker, but that of other neighboring structures as well).

7 Gene Transfer Focusing on Ion Channels

The concept of directly manipulating ion channels to induce pacemaking activity in otherwise ''quiescent'' cardiomyocytes seems, in this context, more appropriate. The two main currents whose activity has been modulated to this end are the outward current of the inward rectifier I_{K1} and the *funny* current, I_f.

Miake et al. [64] created a dominant negative construct that suppresses the I_{K1} current when transfected and co-expressed with the wild-type Kir2.1 into normal cardiomyocytes. Specifically, replacement of amino acids 144–146 in the poreforming H5 region by alanines (Kir2.1AAA) resulted in a dominant negative construct encoding Kir2.1 subunits with non-conducting pores. Kir2.1AAA was then packaged with GFP in a bicistronic adenovirus that would allow expression of both the dominant negative Kir2.1 construct and GFP as a marker to indicate how many and which cells were expressing the two constructs. Following cross clamping of the aorta and the pulmonary artery, adenovirus was injected in the LV cavity of guinea pigs. Myocytes isolated 3–4 days after in vivo transfection with Kir2.1AAA exhibited decreased I_{K1} by 50–90%. Those cells with <80% reduction of I_{K1} exhibited prolongation of APD, decreased RMP and deceleration of Phase 3 of the AP. Moreover, those cells with >80% reduction of I_{K1} exhibited a pacemaker phenotype [64] (Fig. 3). Overall, 40% of the dogs transfected with Kir2.1AAA showed premature beats that were independent of, and faster than, the SAN. The Kir2.1AAA-transduced pacemaker cells were reported to respond to β-AR stimulation with isoproterenol, although no data were presented to this effect. Miake et al. [64] did not explore the source of inward current that mediates pacemaker function, although it is unlikely to be I_f given that, at physiological voltages, I_f is below the limits of measurement in guinea pig LV myocytes [90]. Modeling studies by Silva and Rudy [83] suggest that suppression of I_{K1} on the order of 80% would result in pacemaker activity that is carried primarily by the Na/Ca exchanger (NCX), which depends on levels of intracellular calcium. Furthermore, they found that β-AR responsiveness would strongly depend on the degree of NCX expression, but that a simultaneous β-AR-mediated increase in the slow component of the delayed rectifying current I_{Ks} could limit β-AR sensitivity [83].

Limitations to the studies of Miake et al. [64] included short-term effects and lack of localized expression (there was diffuse injection of Ad-Kir2.1AAA into the coronary circulation of the LV, rather than a localized injection). The investigators did not identify the sites of pacemaker initiation in vivo. Moreover, creating a biopacemaker by attenuating I_{K1} raises the potential for proarrhythmic effects. Specifically, targeting I_{K1} affects repolarization as well as pacemaker function, and the resultant action potential duration (APD) prolongation could enhance transmural dispersion of repolarization that could be arrhythmogenic by itself or, if further prolonged, could provide the basis for development of triggered activity from early afterdepolarizations. These proarrhythmic effects would be even more problematic in the failing heart, where decreased expression and function of I_{K1} can contribute to the development of triggered arrhythmias from altered handling of intracellular calcium [72]. Preliminary theoretical modeling of I_{K1}-downregulated ventricular myocytebased pacemakers revealed significant drawbacks, including inability to drive adjacent, non-pacemaking, cells, as well as lack of protection from electrotonic influences [49]. Nonetheless, given that this approach makes use of native in situ myocytes, it could prove, issues related to the vector excluded, a less immunogenic technique of creating a biological pacemaker. Overall, these experiments open the door for further experimentation on the use of I_{K1} as a modulator of pacemaker activity.

Qu et al. [74] developed a biological pacemaker by adenoviral gene transfer to enhance expression of HCN2, an isoform of the gene that encodes the wild-type pacemaker current I_f that is endogenous to the heart. Injection of adenovirus encoding HCN2 + GFP in the left atrium (LA) appendage resulted in spontaneous LA rhythms

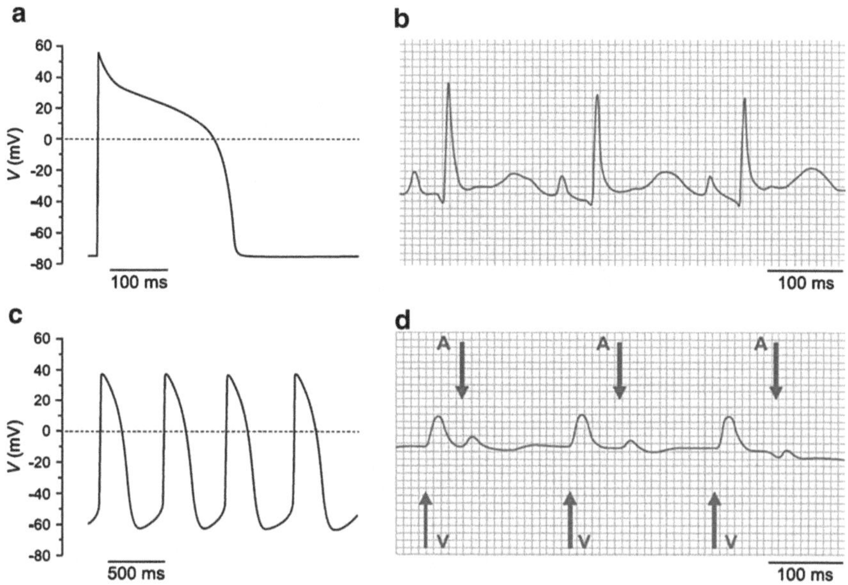

Fig. 3. Effect of I_{K1} suppression in guinea pig ventricular myocytes: **a** Evoked action potentials in control cardiomyocytes. **b** Spontaneous action potentials in I_{K1} suppressed myocytes. c Normal sinus rhythm ECG of control animal at baseline. **d** Dissociated atrial and ventricular rhythms in I_{K1} suppressed animals. P waves (A, *arrow*) and wide QRS complexes (V, *arrow*)on surface ECG. Reprinted from Ref. 64, with permission.

in four of four (vs. 0 of 3 AdGFP-injected dogs) following sinus arrest induced by vagal stimulation. Transfected atrial myocytes exhibited I_f current that was >500 times larger than nonexpressing atrial myocytes. They predicted that the proximal bundle branch system would be the optimal site for injection of HCN2 construct because it would provide organized propagation directly to the ventricle. In a followup study, Plotnikov et al. [71] reported on injection of adenovirus encoding HCN2 into the left bundle branches of dog that were followed for up to 7 days. For the first 48 hours, all dogs that were monitored (whether they received injection of HCN2 + GFP or GFP alone) exhibited multiple premature ventricular complexes or runs of ventricular tachycardia that were attributed to injection-induced local hematoma formation. Following day 2, during vagal stimulation, HCN2-injected dogs exhibited spontaneous rhythms arising from the LV that were faster than those in GFP-injected controls. Isolated LBB tissue studies revealed increased rates of automaticity (Fig. 4) (and greater HCN2 expression) in HCN2-injected dogs, and disaggregated Purkinje fibers from injected regions showed greater I_f magnitude (vs. GFP-injected dogs) (Fig. 5). From published data, it is unclear as to how rapidly biological pacemaker function emerges during vagal stimulation [40, 64]. Moreover, while myocytes overexpressing HCN2 have been shown [76] to respond to β-AR stimulation and increases in cAMP, there were no data to suggest adrenergic responsiveness of the HCN2 construct in vivo. Also, at this vastly preliminary stage, concerns for proarrhythmia stemming from overexpression or variable expression of HCN2 are still to be alleviated. Altogether, though, HCN2 appears an attractive target in the

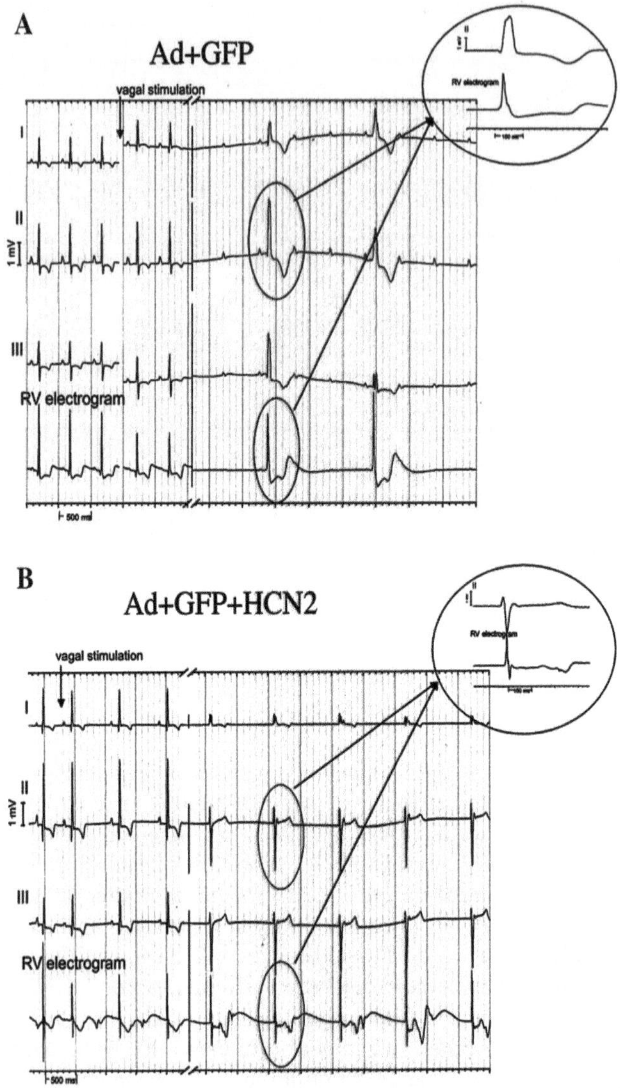

Fig. 4. Occurrence of sustained escape ventricular rhythms in dogs exposed to vagal stimulation following inoculation of their left posterior fascicles with control adenovirus (**a**) or with an adenoviral construct incorporating HCN2 (**b**). Escape rhythm in the latter group of animals is faster, and originates from the left posterior fascicle. Reproduced with permission from Ref. 71

search for the biological pacemaker, mainly because its effects are limited to diastolic depolarization and do not affect repolarization, in contrast to I_{K1}. Further refinement of biological pacemakers will require a number of issues to be further addressed, of which an interesting one would be the interplay of an overexpressed I_f with an underexpressed I_{K1}.

Some insight into the possible combined targeting of I_{K1} and I_f comes from recent work by Kurata et al. [50]. Through nonlinear dynamics modeling and use of the bifurcation theory it is shown, theoretically, that, in the I_{K1}-downregulated human ventricular myocytes (I_{K1} conductance 25% of the control value), expression of I_f facilitates pacemaking. At the same time, expression of I_{st}, and, more imperfectly, of I_{CaT} or I_{CaL}, improves the structural and functional stability to electrotonic loads, reducing the number of pacemaker cells required for pacemaking and driving. The added effects of I_f over-expression were, very interestingly, augmented in a version of the model in which the I_{K1} conductance of the surrounding, non-pacemaking cells, was set to lower levels—as mentioned above, such lower levels of expression and/or function of I_{K1} are to be expected in heart failure. In this case, the expression of I_f itself improved the stability and driving ability of the pacemaker.

8 Ion Channel Proteins That Regulate Pacemaker Ion Channels

The complexity of the sinus node (above) suggests that a biopacemaker based on a single-current gene transfer approach may be inadequate to achieve the degree of sophistication and autonomic regulation of native pacemakers. Further understanding of pacemaker current regulation will be necessary. Along these lines, MinK-related protein (MiRP1), the β subunit for a number of voltagegated potassium channels including the rapid delayed rectifying channel I_{Kr}, has been shown to regulate HCN2 channel expression and gating as well. Qu et al. [75] demonstrated that adenoviral expression of MiRP1 along with HCN2 in cardiac ventricular myocytes resulted in enhanced HCN expression and channel activity. These findings suggest that

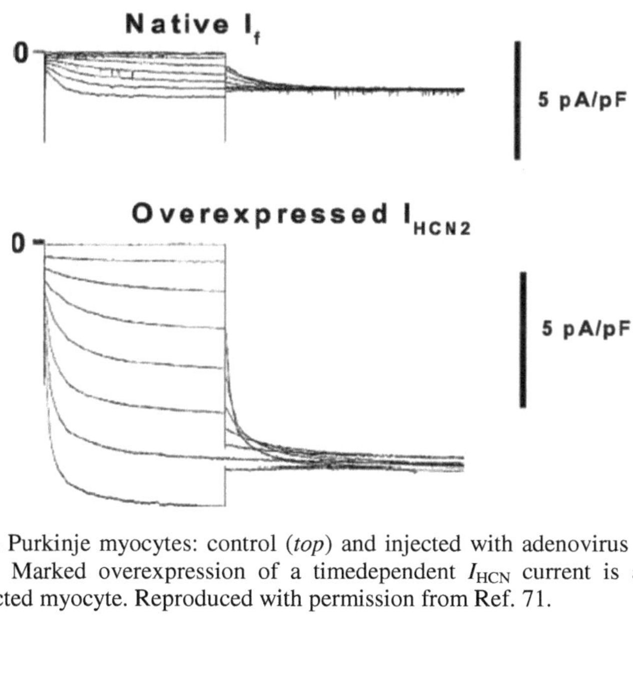

Fig. 5. Canine Purkinje myocytes: control (*top*) and injected with adenovirus carrying HCN2 (*lower panel*). Marked overexpression of a timedependent I_{HCN} current is achieved in the HCN2-transfected myocyte. Reproduced with permission from Ref. 71.

co-expression of regulatory channel proteins along with key pacemaker current protein (such as I_f) may provide a novel approach in which to enhance pacemaker activity and/or decrease the level of pacemaker current protein needed for optimal pacemaker function, and thereby improve the prospects for genebased biopacemakers.

9 Other Requirements for Successful Biopacemakers

Once a candidate gene or set of genes has been established as a valid generator of pacemaking current and is readily available for delivery as needed, there are other important factors, and more or less technical issues, that one would have to consider in the making of a biopacemaker by gene transfer approaches. First, a stable and reliable vector should be available to deliver the construct without unsafe and unwarranted modification of the receiving substrate (such as by inducing immune or inflammatory response, arrhythmias, neo-vascularization producing coronary flow steal, neoplasia, etc). Inflammation and short-term expression are particularly problems with adenoviral gene transfer [45]. Other viral vectors such as adenoassociated virus (AAV) or lentiviruses may offer longer-term expression without significant inflammation, but safety concerns remain. Second, there is a need for long-term duration of pacemaker function. Adenoviral expression typically manifests after 1–2 days, peaking at 3–5 days, and then tending to dissipate over the subsequent 2–3 weeks [28]. AAV could offer more long-term expression, although there are greater limitations to the size of the deliverable construct (compared to adenovirus). AAV-mediated expression of β-galactosidase activity in murine hearts yielded [12] expression and stable enzyme activity lasting at least 6 months. Recent work with AAV type 9 has shown [39] high levels (>80%) of cardiomyocyte transfection following tail vein injection, suggesting that AAV9 may be a robust viral vector for cardiac gene transfer. Third, the pacemaker construct should be energetically independent in vivo or, if dependent upon the receiving tissues for energy, should not become a burden (e.g. create ischemia by a steal phenomenon). Fourth, the biopacemaker should also be able to sustain a wide range of frequencies, ideally through intrinsic properties of the gene construct, preferably mimicking the neuro-humoral chronotropic responsiveness that the native cardiac conduction system displays in response to physiologic or pathophysiologic states (sleep, exercise, pain, fever etc.). This could also be achieved through external pharmacologic modulation, but that situation would be less than ideal, as it would bring with it the risks of polypharmacy in patients, many of whom receive various cardiac or noncardiac medications for other medical conditions. Additionally, the pacemaker should be relatively independent from a wide range of undue pharmacological influences. Fifth, in patients who are still chronotropically competent (i.e., not pacemaker-dependent), the construct should allow higher pacemakers (SAN, for example) to take over the pacemaker role as needed, or to act in synchrony with them. This is supported by recent evidence that permanent pacemaking can have detrimental effects in these patients (especially those with heart failure), and that atrioventricular synchrony is important in the maintenance of a good quality of life and of the left ventricular systolic function [27, 82]. Lastly, another important aspect related to pacemaking is the site from which the impulse is being delivered to the heart. Identification of the best location for the implant is paramount not only for maximizing the obligatory benefit of a backup pacing system, but also for avoiding the risks and outright detrimental effects of single chamber pacing (as single chamber pacemaking can induce significant ventricular dyssynchrony that can adversely affect quality of life and left ventricular systolic function as well as negatively impact on morbidity and mortality) [82, 84]. Such a location would have to be identified for the placement of the implant, to allow it access to the native conduction system. In cases where the conduction system is diseased as well (as in bundle branch disease or trifascicular block), or where cardiac resynchronization is needed, multiple locations should be available and accessible for implant. In the situation where multiple implants are to be delivered, they should be able to sustain their activity in a well-coordinated manner, without crosstalk.

10 Alternatives to Gene Therapy

While further refinements in viral vectors are ongoing, more recent studies have focused on alternative approaches using transplanted fetal and neonatal cardiomyocytes [10, 55], implantation of autologous fibroblasts [9, 47], embryonic stem cells [89] and human mesenchymal stem cells (hMSCs) [73, 86]. Potapova and Plotnikov [73] have used a novel approach of injecting HCN2-transfected hMSCs subepicardially in the canine left ventricle. These hybrid cells incorporated a well-described property of I_f that allows the rate of discharge to accelerate with application of more hyperpolarized voltages [16] (Fig. 2). As such, they displayed automaticity in the relatively hyperpolarized myocardial milieu with which they effectively coupled, and were able to generate rates faster than the ventricular escape rates occurring in the presence of sinus arrest. The notion of using an hMSC genetic construct carrier to deliver a pacemaker current generator to the recipient tissues and protect it thereafter in this new environment (the so called *platform therapy* [78]), is extremely interesting, and presents definite practical advantages. These electrically non-excitable cells [77] are thought to be immunoprivileged, [54] and can be non-immunogenically inoculated with HCN2 via electroporation, with good efficacy. They also contain connexins Cx40 and Cx43, allowing coupling not only with one another, but also with cardiomyocytes [86]. Nonetheless, a better characterization of their properties is called forhMSCs have been used in many studies as a non-homogeneous suspension of cells delivered experimentally to ischemic or failing myocardium. These studies have demonstrated their subsequent differentiation into cardiomyocytes [85], with not only spontaneous electrical activity (at least in cell culture) [92], but also with arrhythmic potential. It has long been thought that hMSC were devoid of any pacemaking current channels or current carriers. More recently, though, $Ca_v1.2$ carrying cardiacspecific I_{CaL} was identified in about 15% of the adult bone marrow human mesenchymal cells [34]. HCN2 was also found in hMSC, but at this point only extremely weak I_f was identified in a very small number of native hMSC studied [34, 74]. It is also unclear what currents would be responsible for the repolarization of HCN-transfected hMSC. This work is discussed further in several articles in this journal.

11 Conclusion

In conclusion, advances in gene transfer have led to a number of potential strategies for creation of a biological pacemaker. Refinements in viral vectors will be paramount, to allow sustained expression and address safety concerns. Moreover, developments with alternative approaches, such as human mesenchymal stem cells, may lead to novel means of establishing stable and physiologically responsive sites of pacemaker activity. While much needs to be done to rival the efficacy of our current electronic pacemakers, the rapid progress in gene transfer and cellular therapy bring us all the more closer to biological pacemakers as supplements (and perhaps, in the future, as replacements) for electronic cardiac pacemakers.

References

1. Anumonwo JM et al (1992) Gap junctional channels in adult mammalian sinus nodal cells. Immunolocalization and electrophysiology. Circ Res 71:229–239
2. Bauer A et al (2004) Inhibitory G protein overexpression provides physiologically relevant heart rate control in persistent atrial fibrillation. Circulation 110:3115–3120
3. Bekeredjian R, Shohet RV (2004) Cardiovascular gene therapy: angiogenesis and beyond. Am J Med Sci 327:139–148
4. Bogdanov KY et al (2006) Membrane potential fluctuations resulting from submembrane Ca2+ releases in rabbit sinoatrial nodal cells impart an exponential phase to the late diastolic depolarization that controls their chronotropic state. Circ Res. DOI 10.1161/01.RES.0000247933.66532.0b
5. Bonke FI (1973) Passive electrical properties of atrial fibers of the rabbit heart. Pflugers Arch 339:1–15
6. Boyett MR et al (1995) Ionic basis of the chronotropic effect of acetylcholine on the rabbit sinoatrial node. Cardiovasc Res 29:867–878
7. Boyett MR, Honjo H, Kodama I (2000) The sinoatrial node, a heterogeneous pacemaker structure. Cardiovasc Res 47:658–687
8. Brown H, DiFrancesco D (1980) Voltageclamp investigations of membrane currents underlying pacemaker activity in rabbit sinoatrial node. J Physiol 308:331–351
9. Bunch TJ et al (2006) Impact of transforming growth factorbeta1 on atrioventricular node conduction modification by injected autologous fibroblasts in the canine heart. Circulation 113:2485–2494
10. Cai J et al (2006) Transplanted neonatal cardiomyocytes as a potential biological pacemaker in pigs with complete atrioventricular block. Transplantation 81:1022–1026
11. Capogrossi MC, Houser SR, Bahinski A, Lakatta EG (1987) Synchronous occurrence of spontaneous localized calcium release from the sarcoplasmic reticulum generates action potentials in rat cardiac ventricular myocytes at normal resting membrane potential. Circ Res 61:498–503
12. Champion HC et al (2003) Robust adenoviral and adenoassociated viral gene transfer to the in vivo murine heart: application to study of phospholamban physiology. Circulation 108:2790–2797
13. De Maziere AM, van Ginneken AC, Wilders R, Jongsma HJ, Bouman LN (1992) Spatial and functional relationship between myocytes and fibroblasts in the rabbit sinoatrial node. J Mol Cell Cardiol 24:567–578

14. DiFrancesco D, Ferroni A, Mazzanti M, Tromba C (1986) Properties of the hyperpolarizing-activated current (if) in cells isolated from the rabbit sinoatrial node. J Physiol 377:61–88
15. DiFrancesco D (1993) Pacemaker mechanisms in cardiac tissue. Annu Rev Physiol 55:455–472
16. DiFrancesco D (2006) Serious workings of the funny current. Prog Biophys Mol Biol 90:13–25
17. Difrancesco D (1991) The contribution of the 'pacemaker' current (if) to generation of spontaneous activity in rabbit sinoatrial node myocytes. J Physiol 434:23–40
18. Difrancesco D (1987) The pacemaker current in the sinus node. Eur Heart J 8(Suppl L): 19–23
19. Donahue JK etal (2000) Focal modification of electrical conduction in the heart by viral gene transfer. Nat Med 6:1395–1398
20. Donahue JK, Bauer A, Kikuchi K, Sasano T (2005) Modification of cellular communication by gene transfer. Ann N Y Acad Sci 1047:157–165
21. Donahue JK, Kikuchi K, Sasano T (2005) Gene therapy for cardiac arrhythmias. Trends Cardiovasc Med 15:219–224
22. Edelberg JM, Aird WC, Rosenberg RD (1998) Enhancement of murine cardiac chronotropy by the molecular transfer of the human beta2 adrenergic receptor cDNA. J Clin Invest 101:337–343
23. Edelberg JM, Huang DT, Josephson ME, Rosenberg RD (2001) Molecular enhancement of porcine cardiac chronotropy. Heart 86:559–562
24. Elmqvist R (1978) Review of early pacemaker development. Pacing Clin Electrophysiol 1:535–536
25. Ertel EA et al (2000) Nomenclature of voltagegated calcium channels. Neuron 25:533–535
26. Fermini B, Nathan RD (1991) Removal of sialic acid alters both T-and L-type calcium currents in cardiac myocytes. Am J Physiol 260:H735–H743
27. Freudenberger RS, Wilson AC, Lawrence-Nelson J, Hare JM, Kostis JB (2005) Permanent pacing is a risk factor for the development of heart failure. Am J Cardiol 95:671–674
28. Gilgenkrantz H et al (1995) Transient expression of genes transferred in vivo into heart using first-generation adenoviral vectors: role of the immune response. Hum Gene Ther 6:1265–1274
29. Gregoratos G et al (2002) ACC/AHA/NASPE 2002 guideline update for implantation of cardiac pacemakers and antiarrhythmia devices: summary article. A report of the American College of Cardiology/American Heart Association Task Force on Practice Guidelines (ACC/AHA/NASPE Committee to Update the 1998 Pacemaker Guidelines). J Cardiovasc Electrophysiol 13:1183–1199
30. Guo J, Ono K, Noma A (1995) A sustained inward current activated at the diastolic potential range in rabbit sinoatrial node cells. J Physiol 483(Pt 1):1–13
31. Guo J, Mitsuiye T, Noma A (1997) The sustained inward current in sinoatrial node cells of guineapig heart. Pflugers Arch 433:390–396
32. Guo J, Noma A (1997) Existence of a low-threshold and sustained inward current in rabbit atrioventricular node cells. Jpn J Physiol 47:355–359
33. Hagiwara N, Irisawa H, Kameyama M (1988) Contribution of two types of calcium currents to the pacemaker potentials of rabbit sinoatrial node cells. J Physiol 395:233–253
34. Heubach JF et al (2004) Electrophysiological properties of human mesenchymal stem cells. J Physiol 554:659–672
35. Hirano Y, Fozzard HA, January CT (1989) Characteristics of L-and T-type Ca2+ currents in canine cardiac Purkinje cells. Am J Physiol 256:H1478–H1492

36. HiranoY, Fozzard HA, January CT (1989) Inactivation properties of T- type calcium current in canine cardiac Purkinje cells. Biophys J 56:1007–1016
37. Holmer SR, Homcy CJ (1991) G proteins in the heart. A redundant and diverse transmembrane signaling network. Circulation 84:1891–1902
38. Hui A et al (1991) Molecular cloning of multiple subtypes of a novel rat brain isoform of the alpha 1 subunit of the voltage-dependent calcium channel. Neuron 7:35–44
39. Inagaki K et al (2006) Robust systemic transduction with AAV9 vectors in mice: efficient global cardiac gene transfer superior to that of AAV8. Mol Ther 14:45–53
40. Inglese J, Freedman NJ, Koch WJ, Lefkowitz RJ (1993) Structure and mechanism of the G protein coupled receptor kinases. J Biol Chem 268:23735–23738
41. Irisawa H, Hagiwara N (1988) Pacemaker mechanism of mammalian sinoatrial node cells.Prog Clin Biol Res 275:33–52
42. Johns DC et al (1995) Adenovirusmediated expression of a voltagegated potassium channel in vitro (rat cardiac myocytes) and in vivo (rat liver). A novel strategy for modifying excitability. J Clin Invest 96:1152–1158
43. Johns DC, Nuss HB, Marban E (1997) Suppression of neuronal and cardiac transient outward currents by viral gene transfer of dominantnegative Kv4.2 constructs. J Biol Chem 272:31598–31603
44. Johns DC, Marban E, Nuss HB (1999) Virusmediated modification of cellular excitability. Ann N Y Acad Sci 868:418–422
45. Jones JM, Wilson KH, Steenbergen C, Koch WJ, Milano CA (2004) Dose dependent effects of cardiac beta2 adrenoceptor gene therapy. J Surg Res 122:113–120
46. Joyner RW,van Capelle FJ (1986) Propagation through electrically coupled cells. How a small SA node drives a large atrium. Biophys J 50:1157–1164
47. Kizana E, Ginn SL, Allen DG, Ross DL, Alexander IE (2005) Fibroblasts can be genetically modified to produce excitable cells capable of electrical coupling. Circulation 111:394–398
48. Kodama I et al (1997) Regional differences in the role of the Ca2+ and Na+ currents in pacemaker activity in the sinoatrial node. Am J Physiol 272:H2793–H2806
49. Kurata Y, Hisatome I, Imanishi S, Shibamoto T (2003) Roles of L-type Ca2+ and delayedrectifier K+ currents in sinoatrial node pacemaking: insights from stability and bifurcation analyses of a mathematical model. Am J Physiol Heart Circ Physiol 285:H2804–H2819
50. Kurata Y, Matsuda H, Hisatome I, Shibamoto T (2006) Effects of pacemaker currents on creation and modulation of human ventricular pacemaker: a theoretical study with application to biological pacemaker engineering. Am J Physiol Heart Circ Physiol
51. Kusumoto FM, Goldschlager N (1996) Cardiac pacing. N Engl J Med 334:89–97
52. Lakatta EG, Maltsev VA, Bogdanov KY, Stern MD, Vinogradova TM (2003) Cyclic variation of intracellular calcium: a critical factor for cardiac pacemaker cell dominance. Circ Res 92:e45–e50
53. Lawrence JH, Johns DC, Chiamvimonvat N, Nuss HB, Marban E (1995) Prospects for genetic manipulation of cardiac excitability. Adv Exp Med Biol 382:41–48
54. Liechty KW et al (2000) Human mesenchymal stem cells engraft and demonstrate sitespecific differentiation after in utero transplantation in sheep. Nat Med 6:1282–1286
55. Lin G et al (2005) Biological pacemaker created by fetal cardiomyocyte transplantation. J Biomed Sci 12:513–519

56. Maltsev VA, Vinogradova TM, Bogdanov KY, Lakatta EG, Stern MD (2004) Diastolic calcium release controls the beating rate of rabbit sinoatrial node cells: numerical modeling of the coupling process. Biophys J 86:2596–2605
57. Maltsev VA, Vinogradova TM, Lakatta EG (2006) The emergence of a general theory of the initiation and strength of the heartbeat. J Pharmacol Sci 100:338–369
58. Mangoni ME et al (2000) Facilitation of the L-type calcium current in rabbit sino atrial cells: effect on cardiac automaticity. Cardiovasc Res 48:375–392
59. Mangoni ME et al (2003) Functional role of L-type Cav1.3 Ca2+ channels in cardiac pacemaker activity. Proc Natl Acad Sci USA 100:5543–5548
60. Mangoni ME et al (2006) Voltage-dependent calcium channels and cardiac pacemaker activity: from ionic currents to genes. Prog Biophys Mol Biol 90:38–63
61. Marionneau C et al (2005) Specific pattern of ionic channel gene expression associated with pacemaker activity in the mouse heart. J Physiol 562:223–234
62. Matsuura H, Ehara T, Ding WG, OmatsuKanbe M, Isono T (2002) Rapidly and slowly activating components of delayed rectifier K(+) current in guineapig sino atrial node pacemaker cells. J Physiol 540:815–830
63. Matthes J et al (2004) Disturbed atrioventricular conduction and normal contractile function in isolated hearts from Cav1.3-knockout mice. Naunyn Schmiedebergs Arch Pharmacol 369:554–562
64. Miake J, Marban E, Nuss HB (2002) Biological pacemaker created by gene transfer. Nature 419:132–133
65. Mitsuiye T, Shinagawa Y, Noma A (2000) Sustained inward current during pacemaker depolarization in mammalian sinoatrial node cells. Circ Res 87:88–91
66. Noma A, Morad M, Irisawa H (1983) Does the ''pacemaker current'' generate the diastolic depolarization in the rabbit SA node cells? Pflugers Arch 397:190–194
67. Nuss HB et al (1996) Reversal of potassium channel deficiency in cells from failing hearts by adenoviral gene transfer: a prototype for gene therapy for disorders of cardiac excitability and contractility. Gene Ther 3:900–912
68. Nuss HB, Marban E, Johns DC (1999) Overexpression of a human potassium channel suppresses cardiac hyperexcitability in rabbit ventricular myocytes. J Clin Invest 103:889–896
69. Oosthoek PW et al (1993) Immunohistochemical delineation of the conduction system. I: The sinoatrial node. Circ Res 73:473–481
70. Platzer J et al (2000) Congenital deafness and sinoatrial node dysfunction in mice lacking class D L-type Ca2+ channels. Cell 102:89–97
71. Plotnikov AN et al (2004) Biological pacemaker implanted in canine left bundle branch provides ventricular escape rhythms that have physiologically acceptable rates. Circulation 109:506–512
72. Pogwizd SM, Schlotthauer K, Li L, Yuan W, Bers DM (2001) Arrhythmogenesis and contractile dysfunction in heart failure: roles of sodium–calcium exchange, inward rectifier potassium current, and residual beta-adrenergic responsiveness. Circ Res 88:1159–1167
73. Potapova I et al (2004) Human mesenchymal stem cells as a gene delivery system to create cardiac pacemakers. Circ Res 94:952–959
74. Qu J et al (2003) Expression and function of a biological pacemaker in canine heart. Circulation 107:1106–1109
75. Qu J et al (2004) MiRP1 modulates HCN2 channel expression and gating in cardiac myocytes. J Biol Chem 279:43497– 43502
76. Qu J et al (2001) HCN2 overexpression in newborn and adult ventricular myocytes: distinct effects on gating and excitability. Circ Res 89:E8–E14

77. Ravens U (2006) Electrophysiological properties ofstem cells. Herz 31:123–126
78. Rosen MR (2005) 15th annual Gordon K. Moe Lecture. Biological pacemaking: in our lifetime? Heart Rhythm 2:418–428
79. Satoh H (2003) Sinoatrial nodal cells of mammalian hearts: ionic currents and gene expression of pacemaker ionic channels. J Smooth Muscle Res 39:175–193
80. Satoh H, Tsuchida K (1993) Comparison of a calcium antagonist, CD-349, with nifedipine, diltiazem, and verapamil in rabbit spontaneously beating sinoatrial node cells. J Cardiovasc Pharmacol 21:685–692
81. Shi W et al (1999) Distribution and prevalence of hyperpolarization-activated cation channel (HCN) mRNA expression in cardiac tissues. Circ Res 85:e1–e6
82. Shukla HH et al (2005) Heart failure hospitalization is more common in pacemaker patients with sinus node dysfunction and a prolonged paced QRS duration. Heart Rhythm 2:245–251
83. Silva J, Rudy Y (2003) Mechanism of pacemaking in I(K1)-downregulated myocytes. Circ Res 92:261–263
84. Sweeney MO, Hellkamp AS, Lee KL, Lamas GA (2005) Association of prolonged QRS duration with death in a clinical trial of pacemaker therapy for sinus node dysfunction. Circulation 111:2418–2423
85. Toma C, Pittenger MF, Cahill KS, Byrne BJ, Kessler PD (2002) Human mesenchymal stem cells differentiate to a cardiomyocyte phenotype in the adult murine heart. Circulation 105:93–98
86. Valiunas V et al (2004) Human mesenchymal stem cells make cardiac connexins and form functional gap junctions. J Physiol 555:617–626
87. Verheijck EE, van Ginneken AC, Wilders R, Bouman LN (1999) Contribution of L-type Ca2+ current to electrical activity in sinoatrial nodal myocytes of rabbits. Am J Physiol 276:H1064–H1077
88. Vinogradova TM, Maltsev VA, Bogdanov KY, Lyashkov AE, Lakatta EG (2005) Rhythmic Ca2+ oscillations drive sinoatrial nodal cell pacemaker function to make the heart tick. Ann N Y Acad Sci 1047:138–156
89. Xue T et al (2005) Functional integration of electrically active cardiac derivatives from genetically engineered human embryonic stem cells with quiescent recipient ventricular cardiomyocytes: insights into the development of cell-based pacemakers. Circulation 111:11–20
90. Yu H, Chang F, Cohen IS (1993) Phosphatase inhibition by calyculin A increases i(f) in canine Purkinje fibers and myocytes. Pflugers Arch 422:614–616
91. Zhang Z et al (2002) Functional Roles of Ca(v)1.3 (alpha(1D)) calcium channel in sinoatrial nodes: insight gained using genetargeted null mutant mice. Circ Res 90: 981–987
92. Zhang YM, Hartzell C, Narlow M, Dudley SC Jr (2002) Stem cell-derived cardiomyocytes demonstrate arrhythmic potential. Circulation 106:1294–1299

Biological Pacemakers Based on I_f

Michael R. Rosen[1,2,3], Peter R. Brink[3], Ira S. Cohen[1,3], and Richard B. Robinson[1]

[1] Department of Pharmacology, Center for Molecular Therapeutics,
College of Physicians and Surgeons of Columbia University,
630 West 168 Street, PH7West-321,
New York, NY 10032, USA
mrr1@columbia.edu
[2] Department of Pediatrics, Columbia University,
New York, NY, USA
[3] Departments of Physiology and Biophysics,
Institute of Molecular Cardiology, SUNY Stony Brook,
Stony Brook, NY, USA

Abstract. Biological pacemaking as a replacement for or adjunct to electronic pacemakers has been a subject of interest since the 1990s. In the following pages, we discuss the rational for and progress made using a hyperpolarization activated, cyclic nucleotide gated channel isoform to carry the I_f pacemaker current in gene and cell therapy approaches. Both strategies have resulted in effective biological pacemaker function over a period of weeks in intact animals. Moreover, the use of adult human mesenchymal stem cells as a platform for carrying pacemaker genes has resulted in the formation of functional gap junctions with cardiac myocytes in situ leading to effective and persistent propagation of pacemaker current. The approaches described are encouraging, suggesting that biological pacemakers based on this strategy can be brought to clinical testing.

Keywords: HCN isoforms, Electronic pacemakers, Sinoatrial node, Atrioventricular block, Gene therapy, Cell therapy.

1 Introduction

Agreeing to prepare this manuscript implies acceptance of the proposition that biological pacemakers are a need of the community. Yet, this is less than a tested and proven proposition. In fact, given the availability of superb electronic pacemaker technology, it might be argued that biological pacemaking should be low on the list of priorities for new therapeutic ventures. However, there are compelling reasons for proceeding in this direction [3, 12, 13]. In part, the reasons are therapeutic: as good as electronic pacemakers are, they remain palliatives; to develop a biological pacemaker means to embark on discovering a cure. In part, the reasons are educational: because electronic pacemakers are as good as they are, there is no need for a biological pacemaker today, tomorrow is soon enough. The key is that sufficient time can and must be taken to

(1) learn what is needed to get the method right, and
(2) to be certain that the therapy is as good as it is expected to be (and superior to electronic pacemaking) before bringing it to the clinic.

J.A.E. Spaan (Ed.): Biopacemaking, BIOMED, pp. 63–78, 2007.
springerlink.com

This learning process extends far beyond biological pacemaking per se; it can be applied to many areas of gene and cell therapy. In other words, rather than the lamentable race "to be first" that too often complicates medical therapies these days, the intent should be to provide the right therapy at the earliest possible date.

There are many approaches to preparing biological pacemakers, all of which are reviewed in this volume. One or more should find its way to the clinic. In considering all approaches, we should remember that morphologically, the sinus node is a complex structure in which different types of nodal cells are present, all of them embedded in collagen. In addition, sinoatrial nodal cells are heterogeneous in terms of connexin expression, and there is a clear cell size-dependence in the pattern of connexin expression. With regard to this complexity, one approach that might be taken would be to engineer a morphological and functional replica of the sinus node. Rather than assume this daunting task, we have taken a lesson from our engineering colleagues who designed the electronic pacemaker; that is, we are working to finetune a structure that mimics the sinus node functionally without recapitulating it morphologically.

This approach to biological pacemaking revolves around both gene and cell therapies [5,8,10], with the focus on one particular target, the HCN (hyperpolarization activated, cyclic nucleotide gated) gene isoforms responsible for the I_f pacemaker current [1]. We have chosen the HCN isoforms for two reasons: first, because together they constitute the family that initiates pacemaker activity in the mammalian heart (with HCN4 and HCN1 predominating in sinoatrial node and HCN2 in ventricular specialized conducting system); and second, because they not only initiate pacemaker activity, but their activation is sped by catecholamines and slowed by acetylcholine, making them autonomically responsive. And autonomic responsiveness is and should be a cornerstone of pacemaker activity in heart: lack of this is a key shortcoming of electronic pacemakers.

To date, we have used HCN2 as our primary research tool in a two-pronged approach. A gene therapy limb [5, 10] utilizes adenoviral vectors to test mutant as well as chimeric genes in an effort to optimize pacemaker activity and to test interactions between biological and electronic pacemaker therapies (so called tandem pacemaking). Importantly, the replication-deficient adenovirus we have used results in only transient (about 2 weeks) expression of pacemaker function. While not promising for any long-term therapeutic modality, this approach does provide a convenient means for proof-of-concept experiments. We are working as well with adeno-associated virus to enable more durable expression: these experiments are still in their early phases.

The second limb of our research, that of cell therapy, uses adult human mesenchymal stem cells (hMSCs) as a platform for delivery of HCN constructs to the myocardium [8]. Both the gene and cell therapy approaches will be reviewed here.

2 Gene Therapy Using HCN

Figure 1 provides a starting point for understanding the role of HCN genes and the I_f current they carry in initiating the pacemaker potential. It also explains why we have focused on the HCN family. In brief, phase 4 depolarization is initiated by inward sodium current activated on hyperpolarization of the cell membrane and is continued and sustained by other currents [1]. The latter incorporate a balance between inward currents carried by calcium and the sodium/calcium exchanger and outward currents carried by potassium. Activation of the pacemaker potential is increased by betaadrenergic catecholamines and reduced by acetylcholine through their respective G protein-coupled receptors and the adenylyl cyclase-cyclic AMP second messenger system via a cyclic AMP binding site near the carboxyl terminus of the channel.

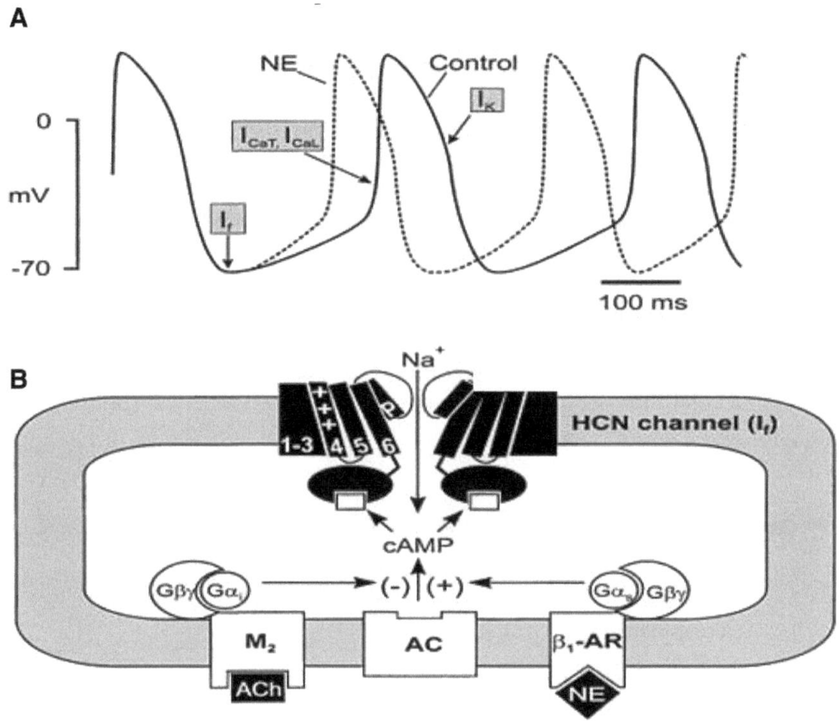

Fig. 1. The role of I_f in generation of pacemaker potentials in the sinoatrial node (SAN). **a** Pacemaker potentials in the SAN under control conditions, and after β-adrenergic stimulation with norepinephrine. The four major currents that control the generation of the pacemaker potential are indicated: I_f [produced by hyperpolarization activated, cyclic nucleotide gated (HCN) channels], T-type (I_{CaT}) and L-type (I_{CaL}) calcium currents, and repolarizing K currents (I_K). **b** Scheme of an SAN cell showing the regulation of the HCN channel by up-or downregulation of cellular cyclic adenosine monophosphate (cAMP). *M2* type-2 muscarinic receptor, *ACh* acetylcholine, *AC* adenylyl cyclase, Gαi G-protein α subunit (inhibits AC), *Gβγ* G-protein *βγ* subunit, *β*1-AR *β*1-adrenergic receptor, *G*αs G-protein a subunit (stimulates AC), ΔV shift of the voltage dependence of HCN channel activation induced by increase or decrease of cAMP (reprinted by permission from reference [1]).

We reasoned that overexpression of I_f in either secondary pacemaker tissues of the cardiac specialized conducting system or in non-pacemaker cells of the myocardium could provide a nidus of pacemaker activity to drive the heart in a ‘‘demand’’ mode in the absence of dominant pacemaker function of the sinus node or failure of impulse propagation via the atrioventricular node. We chose HCN2 as the gene of interest because its kinetics is more favorable than those of HCN4 and its cyclic adenosine monophosphate responsiveness is greater than that of HCN1. Initial experiments were performed in neonatal rat myocytes in culture and indicated that not only could an overexpressed pacemaker current increase beating rate, but that mutations on the HCN2 pacemaker gene and/or the addition of appropriate accessory channel subunits could modify the characteristics of the expressed current in a manner that might be expected to further enhance the beating rate [2, 6, 9, 11]. These neonatal ventricular myocytes manifest a small endogenous pacemaker current (Fig. 2, top left), and when infected with an adenovirus carrying HCN2, express a markedly larger pacemaker current (Fig. 2, bottom left). When we compare the spontaneous beating rate of monolayer cultures infected with either a control virus incorporating green fluorescent protein (GFP) as a control and marker, or expressing the GFP and HCN2 genes, the HCN2/GFP-expressing cultures beat significantly faster (Fig. 2, right).

One approach to further enhance pacemaker activity is increasing the magnitude of the expressed current and/or speed its kinetics of activation. As illustrated in Fig. 3, both of these goals can be achieved by co-expressing HCN2 with its beta subunit, MiRP1. In this experiment, the myocyte cultures were infected with the HCN2 adenovirus and a second virus that was a vehicle for either GFP or an HA-tagged form of MiRP1. The result was a significant increase in current magnitude (top panels) and acceleration of activation and deactivation kinetics (bottom panels)

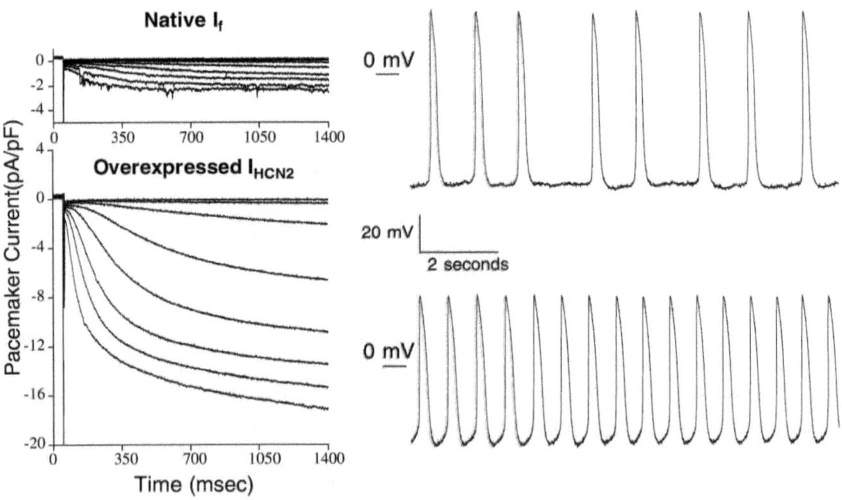

Fig. 2. Hyperpolarization activated, cyclic nucleotide gated (HCN) overexpression in neonatal rat ventricular myocyte culture increases spontaneous rate. *Left panel* Representative current traces of endogenous pacemaker current from a control myocyte expressing only green fluorescent protein (GFP) (*top*) and from a myocyte overexpressing murine HCN2 (*bottom*). *Right panel* Representative recordings of spontaneous action potentials from monolayer cultures infected with a GFP expressing adenovirus (*top*) and an HCN2 expressing adenovirus (*bottom*) (reprinted by permission from reference [9]).

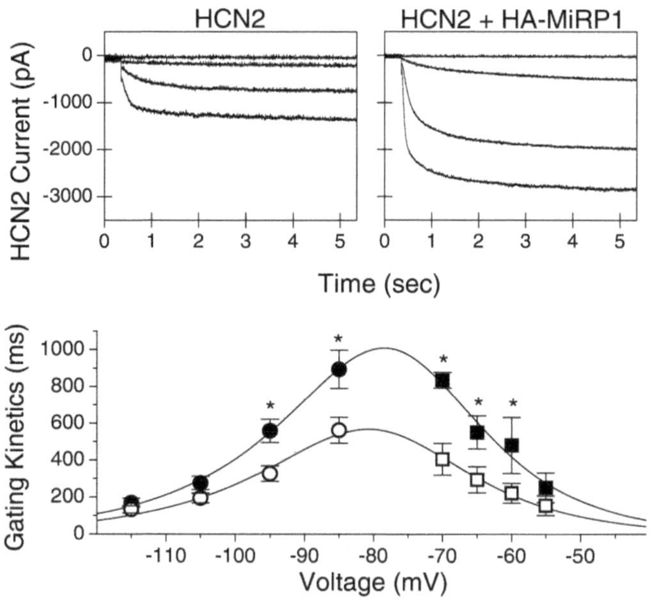

Fig. 3. Co-expression of [hyperpolarization activated, cyclic nucleotide gated (HCN)] HCN2 and HA-tagged MiRP1 in neonatal ventricular myocytes alters current magnitude and kinetics. *Top* Representative HCN2 currents recorded from a myocyte co-infected with adenoviruses expressing HCN2 and green fluorescent protein (GFP) (as a control; *left*) and one co-infected with adenoviruses expressing HCN2 and HA-tagged MiRP1 (*right*). Test voltages were –45, –65, –85, –105 mV; the holding potential was –35 mV. Bottom Effect of HA-tagged MiRP1 co-expression on kinetics of HCN2 activation and deactivation. Activation (*circles*) and deactivation (*squares*) time constants are plotted as a function of voltage for HCN2 and MiRP1 co-expression (*unfilled symbols*) and compared to those for HCN2 and GFP co-expression (*filled symbols*). The solid lines are the best-fit curves to the equation $\tau = 1/(A1\times\exp(-V/B1) + A2\times\cdot\exp(V/B2))$ where τ is the activation or deactivation kinetic time constant; $A1$, $A2$, $B1$, and $B2$ are calculated fitting parameters. *Asterisk* indicates significant differences (reprinted by permission from reference [11]).

[11]. In other experiments, we have explored a point mutation in murine HCN2 (E324A) that was reported to exhibit both faster kinetics and a more positive activation relation than HCN2 [2], both characteristics that should enhance pacemaking. In preliminary data, we found that while the mutation did indeed show these preferred characteristics in myocyte cultures, it also expressed less well than the wild-type HCN2 gene, resulting in significantly reduced current magnitude [6]. Finally the HCN2 construct was responsive both to the acceleratory effect of beta-adrenergic catecholamine and to the deceleratory effects of acetylcholine.

Encouraged by the implications of the cell culture work, we then proceeded to test proof-of-concept by injecting a small quantity of HCN2 and GFP genes in an adenoviral vector into canine left atrium [10]. We permitted the animals to recover

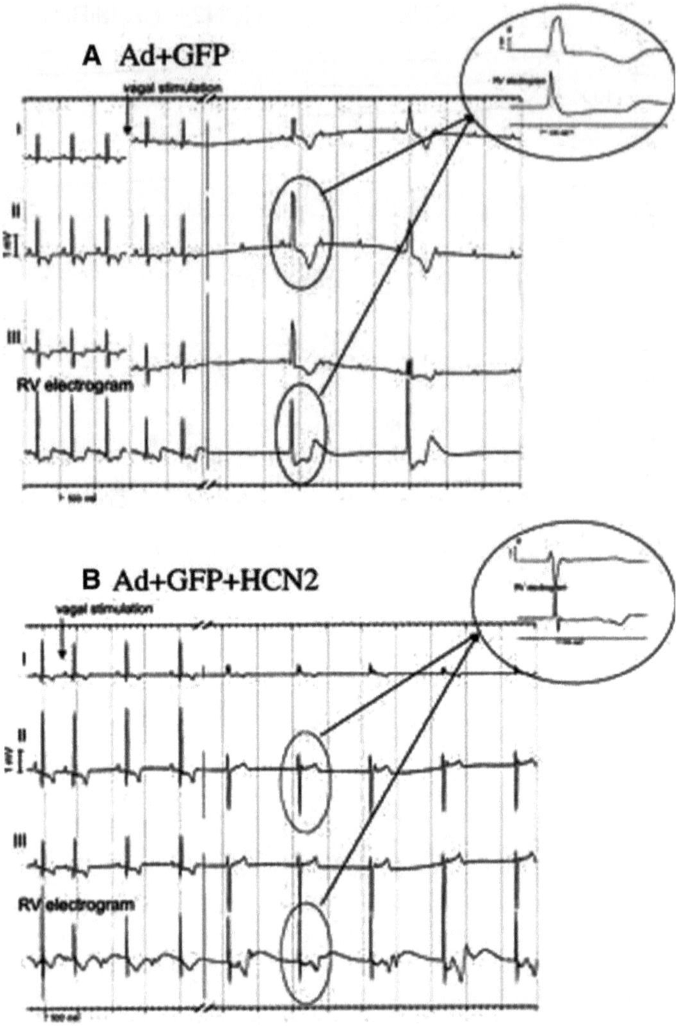

Fig. 4. Representative experiments with AdGFP injection (**a**) and AdGFP + AdHCN2 injection (**b**) into the left bundle branch. Note that, at the outset (*left side of each panel*), sinus rhythm of comparable rate occurs in both animals. In both dogs, the primary effect of vagal stimulation (onset at *arrow*) was to induce AV block. This was followed by slow idioventricular rhythm in the AdGFP-injected dog (**a**; interval between left and right traces lasted 22 s, near outset of which vagal stimulation was initiated). In the AdHCN2-injected dog (**b**), vagal stimulation was followed by far more rapid idioventricular rhythm. Interval between left and right traces was of 5-s duration. Insets are magnifications of lead II and RV electrogram impulses indicated in basic traces. These show that in the AdGFP-injected animal, electrogram initiation was early in the QRS complex, whereas in the AdHCN2-injected animal, RV electrogram initiation occurred later (reprinted by permission from reference [5]).

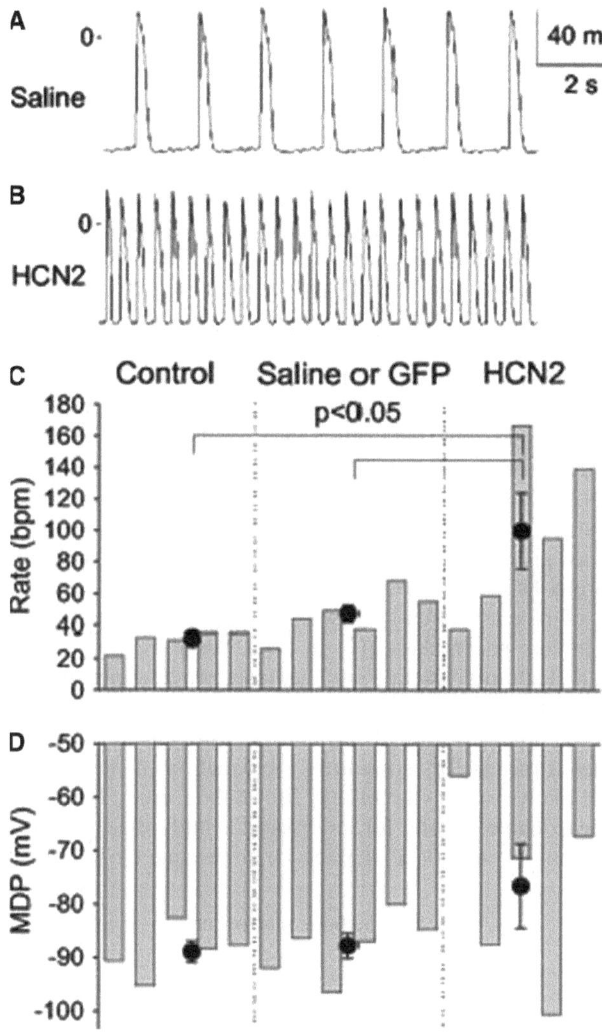

Fig. 5. Automatic rate and maximum diastolic potential (MDP) in left bundle branch preparations. **a** and **b** Representative recordings from 1 saline- and 1 AdHCN2-injected dog. Rate (**c**) and MDP (**d**) for individual dogs, with *black circles* indicating mean ± SEM for each group (reprinted by permission from reference [5]).

and several days later stimulated the right vagus nerve to induce sinoatrial slowing and/or block. In this setting, pacemaker activity originated in the left atrium and was pacemapped to the site of injection. Increasing the intensity of the vagal stimulation and adding left vagal stimulation caused cessation of biological pacemaker activity, consistent with parasympathetic responsiveness. We disaggregated atrial myocytes from the site of injection, demonstrated overexpressed pacemaker current, and interpreted the sum total of results as indicating that such overexpressed pacemaker current could provide escape beats under circumstances of sinus slowing [10].

The next steps involved catheter injection of the same construct into the canine proximal left ventricular conducting system, under fluoroscopic control [5]. Animals so injected demonstrated idioventricular rhythms having rates of 50–60 bpm when sinus rhythm was suppressed by vagal stimulation. For the HCN2 group, the rhythms mapped to the site of injection (Fig. 4). When bundle branch tissues were removed from the heart and studied with microelectrodes, we found that automaticity in those injected with HCN2 exceeded that in control preparations. As shown in Fig. 5, there was a significantly greater spontaneous rate generated by the HCN2 injected bundle branches than by those injected with either saline or virus carrying GFP, alone.

Our most recent work has expanded in two areas as reported in preliminary fashion [6]. First, we have tested the possibility that injecting an adenovirus carrying the E324A mutant described above might provide an effective alternative to HCN2 in experiments in vivo. While the demonstration of favorable activation kinetics in situ suggested E324A-based pacemakers might increase pacemaker rate, we were concerned the lesser magnitude of current expression might offset the potential benefit. We also believed that regardless of its effects on basal rate, E324A might bring about a greater sympathetic response than HCN2. We found that E324A-infected dogs manifested basal rates that did not differ significantly from those of HCN2-in-fected animals, while their catecholamine-responsive-ness was greater. As such, the E324A mutant represents only a subtle variation on the parent, HCN2, construct.

The other area of recent interest has been the exploration of tandem therapy with electronic and biological pacemakers [6]. In brief, we have found that using the two together provides an electronic baseline rate that insures a heartbeat even if the biological component fails, while the biological component provides the autonomic rate responsiveness so important to normal physiologic function. In addition, the electronic component provides a means for monitoring the function of the biological pacemaker while the latter will likely prolong the battery life of the former. Hence, the two together should provide a reasonable pairing for any phase 1 study that may be considered and may provide significant therapeutic advantage for some time thereafter.

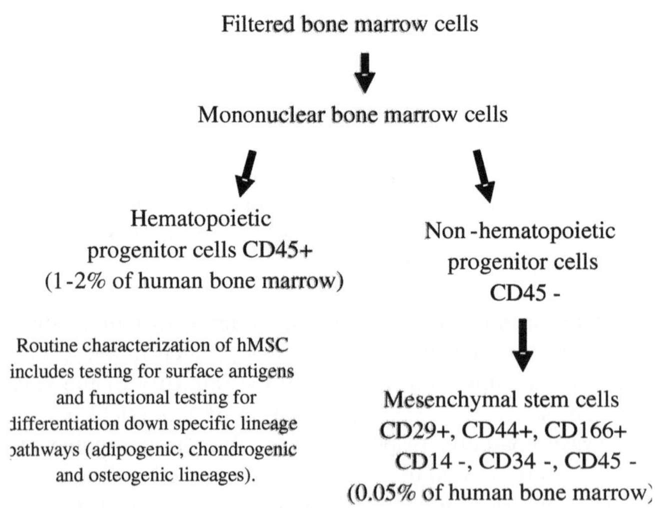

Fig. 6. Origin of adult human mesenchymal stem cells

3 Cell Therapy with hMSCs

Use of hMSCs brought into play the concept of a platform therapy. We viewed the hMSCs as a favorable platform candidate as the literature suggested they might be immunoprivileged [4] and as such would hopefully not give rise to a rejection

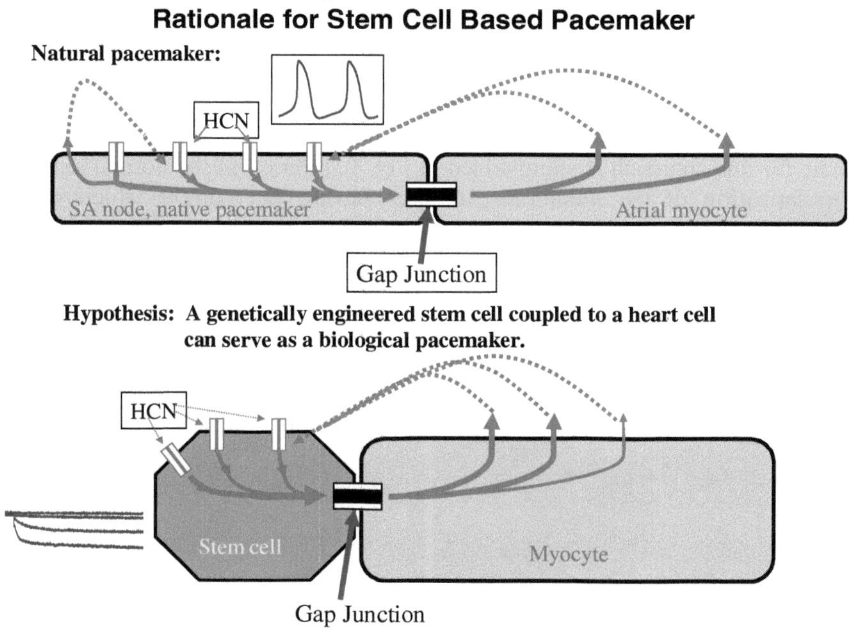

Fig. 7. Initiation of spontaneous rhythms by wild-type or genetically engineered pacemaker cells as well as by genetically engineered stem cell pacemakers. *Top panel* In a native pacemaker cell or in a myocyte engineered to incorporate pacemaker current via gene transfer, action potentials (*inset*) are initiated via inward current flowing through transmembrane hyperpolarization activated, cyclic nucleotide gated (HCN) channels. These open when the membrane repolarizes to its maximum diastolic potential and close when the membrane has depolarized during the action potential. Current flowing via gap junctions to adjacent myocytes results in their excitation and the propagation of impulses through the conducting system. *Bottom panel* A stem cell has been engineered to incorporate HCN channels in its membrane. These channels can only open, and current can only flow through them (*inset*) when the membrane is hyperpolarized; such hyperpolarization can only be delivered if an adjacent myocyte is tightly coupled to the stem cell via gap junctions. In the presence of such coupling and the opening of the HCN channels to induce local current flow, the adjacent myocyte will be excited and initiate an action potential that then propagates through the conducting system. The depolarization of the action potential will result in the closing of the HCN channels until the next repolarization restores a high negative membrane potential. In summary, wild-type and genetically engineered pacemaker cells incorporate in each cell all the machinery needed to initiate and propagate action potentials. In contrast, in the stem cell–myocyte pairing, two cells together work as a single functional unit whose operation is critically dependent on the gap junctions that form between the two disparate cell types (reprinted by permission from reference [13]).

response. This was important because we believe that in the tradeoff between biological and electronic pacemakers, any need for immunosuppression using the former approach would render it clinically undesirable. hMSCs are obtained readily commercially or from the bone marrow and are identified by the presence of CD44 and CD29 surface markers as well as by the absence of other markers (Fig. 6) that are specific for hematopoietic progenitor cells.

Using a gene chip analysis, we determined that the hMSCs do not carry message for HCN isoforms; importantly, they do have a significant message level for the gap junctional protein, connexin 43. This is critical because the theory behind platform therapy was that the hMSC would be loaded with the gene of interest (HCN2 in this case) and implanted into myocardium [13]. However, having a cell loaded with a signal would not work unless the cell formed functional connections with its neighbors. The rationale for this approach is summarized in Fig. 7. In brief, in the normal sinus node, hyperpolarization of the membrane initiates inward (I_f) current which generates

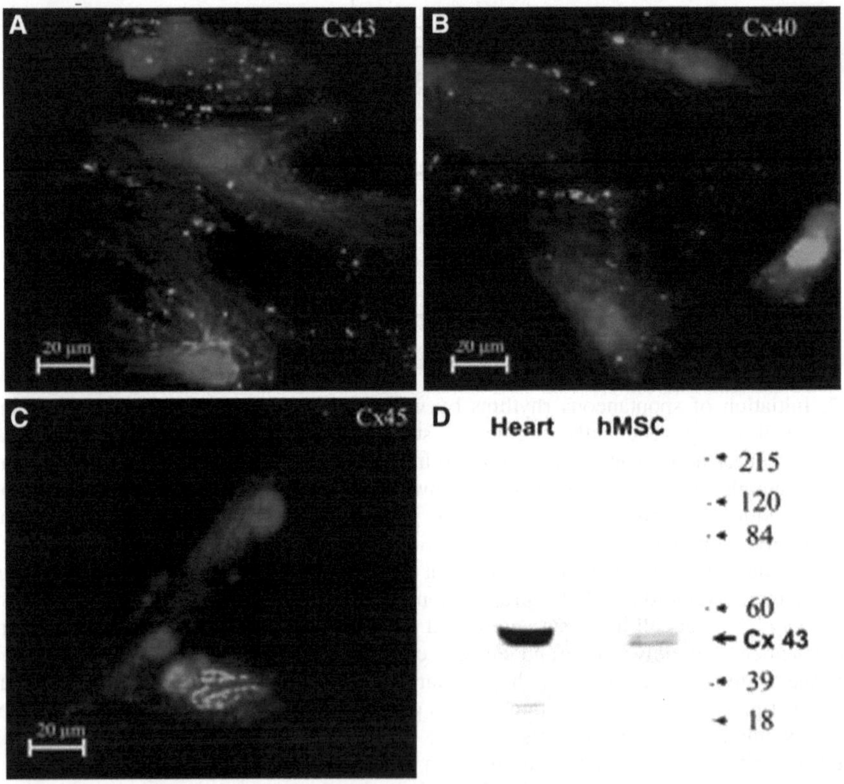

Fig. 8. Identification of connexins in gap junctions of human mesenchymal stem cells (hMSCs). Immunostaining of Cx43 (**a**), Cx40 (**b**), and Cx45 (**c**). **d** Immunoblot analysis of Cx43 in canine ventricular myocytes and hMSCs. Migration of molecular weight markers is indicated to the *right* of the blot (reprinted by permission from reference [14]).

phase 4 depolarization and an automatic rhythm. The changes in membrane potential result in current flow via the low resistance gap junctions such that the action potential propagates from one cell to the next. Our view of the hMSC as a platform was that it would be loaded with the HCN2 gene via electroporation, thereby avoiding any viral component in the process [3, 8, 12, 13]. The hMSC would have to be coupled effectively to the adjacent myocyte. If this occurred, then the high negative membrane potential of coupled myocytes would hyperpolarize the hMSC, opening the HCN channels and permitting inward current to flow. This current, in turn, would propagate through the low resistance gap junctions, depolarize a coupled myocyte, and bring it to threshold potential, resulting in an action potential that would then propagate further in the conducting system. In other words, the hMSC and the myocyte each would have to carry an essential piece of machinery: the myocyte would bring the ionic components that generate an action potential, the hMSC would carry the pacemaker current, and—if gap junctions were present—the two separate structural entities would function as a single, seamless physiologic unit.

A key question then was: are gap junctions formed between hMSCs and myocytes? The answer is yes, as is shown in Fig. 8. Note, using immunostaining connexins 43 and 40 are clearly demonstrable, while connexin 45 is not. Western blots also showed the presence of both connexins 40 and 43. Moreover, injection of current into an hMSC in close proximity to a myocyte results in current flow to the myocyte (Fig. 9) [14], clearly indicating the existence of gap junctions and the occurrence of electrical coupling. Another important question here is how critical the extent of coupling between the engineered nodal cells and surrounding myocardium may be.

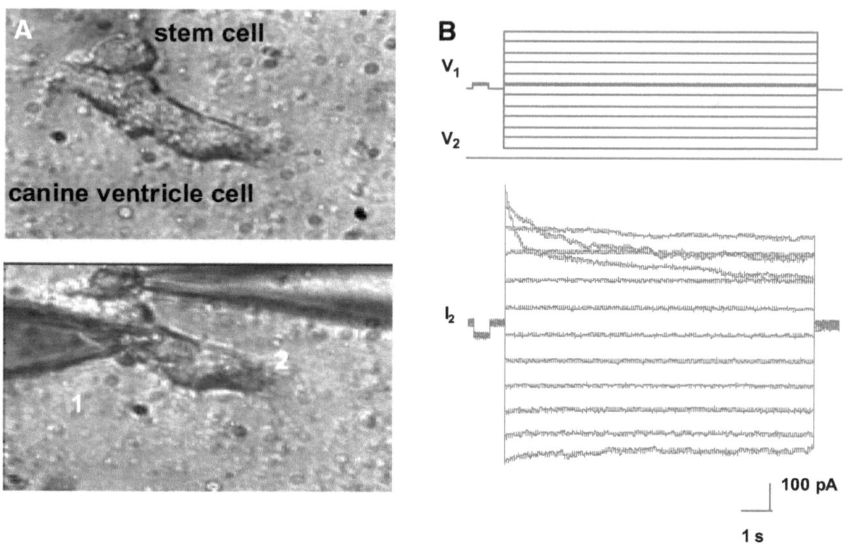

Fig. 9. Current flow demonstrating coupling between a human mesenchymal stem cell (hMSC) and a canine myocyte. **a** Phase-contrast micrograph of a hMSC–canine ventricular myocyte pair before (*upper*) and after (*lower*) impaling each with a micro- electrode. *Panel b* Monopolar pulse protocol (*V*1 and *V*2) and associated macroscopic junctional currents (*I*) exhibiting asymmetrical voltage dependence (reprinted by permission from reference [14]).

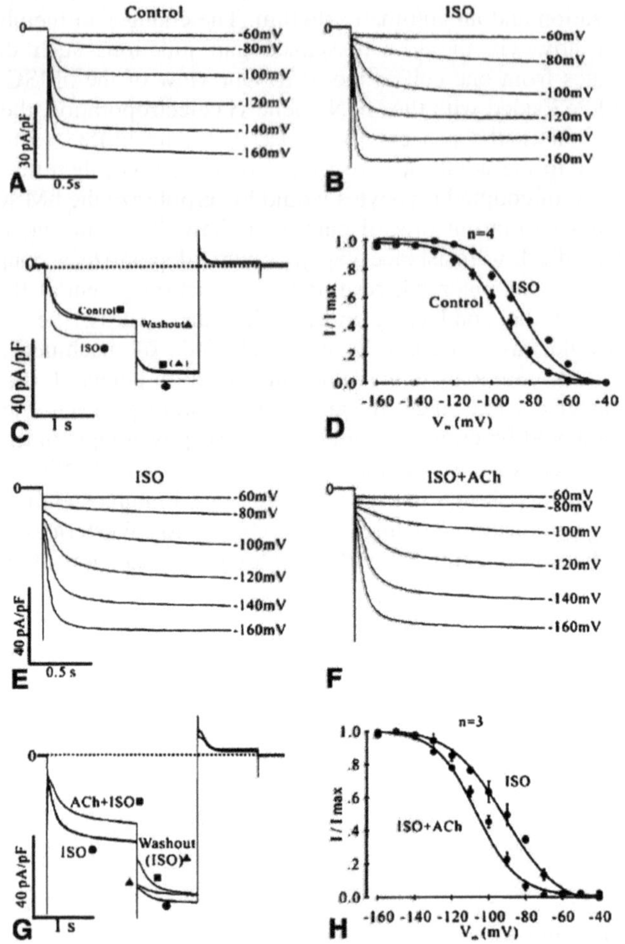

Fig. 10. Modulation of I_f activation by isoproterenol (ISO, *panels a–d*) and acetylcholine (ACh, *panels e–h*) in human mesenchymal stem cells transfected with the mHCN2 gene. I_f activation in the absence (**a**) and presence of ISO, 1×10^{-6}mol/l (**b**). The currents at –80 and –100 mV in isoproterenol are larger than those in control, whereas the currents in both conditions are almost equal at –160 mV. **c** Voltage dependence of activation of I_f in control, ISO, and washout using a two-step pulse protocol. **d** Boltzmann fit to the normalized density of tail currents to construct activation curve. Two-pulse protocol was initiated from a holding potential of –30 mV. The first step was to –100 mV for 1.5 s followed by a second step to –150 mV for 1 s. Voltage was then stepped to +15 mV for 1 s to rapidly deactivate the current and then returned to the holding potential. *Panels e–h* Modulation of I_f activation by acetylcholine (ACh) in the presence of ISO. I_f activation in the presence of ISO and in the absence (**e**) and presence (**f**) of ACh (1×10^{-6} mol/l). Addition of acetylcholine reduces the membrane currents. However, they are almost identical at –160 mV, consistent with a negative shift in activation induced by acetylcholine. **g** Same two-step protocol as in *panel c*, for ISO (1×10^{-6}mol/l) alone and ISO + ACh. h Boltzmann fit to normalized currents. Activation curve was constructed with the same protocol as in *panel d* (reprinted by permission from reference [8]).

One might expect that if coupling is too tight, nodal cells will be clamped by surrounding myocardial cells to their resting membrane potential, thus suppressing spontaneous activity. This is an issue we have not addressed, as we found that in our initial experiments, the ratio between current developed, cell–cell coupling and propagation of impulses was such that pacemaker expression was readily and consistently apparent. This may suggest that our experiments exist on a saturated part of a dose–response curve between cells delivered, current expressed and gap junctions formed. Future experiments will test this hypothesis.

The next issue we considered was the autonomic responsiveness of the hMSCs [8]. As shown in Fig. 10, panels a–e, the addition of isoproterenol to hMSCs loaded with HCN2 resulted in a shift in activation such that increased current flowed at more positive potentials. The result, as would be expected for native HCN2, should be an increased pacemaker rate. In Fig. 10, panels e–h show the response of I_f expressed by hMSCs to acetylcholine. Acetylcholine alone had no effect on current, but in the presence of isoproterenol, antagonized the beta-adrenergic effect of the latter. This is entirely consistent with the physiologic phenomenon of accentuated antagonism.

We then injected hMSCs loaded with HCN2 into the hearts of dogs in which vagal stimulation was used to terminate sinoatrial pacemaker function and/or atrioventricular conduction [11]. This was done effectively and demonstrated spontaneous pacemaker function that was pace-mapped to the site of injection (Fig. 11). Moreover, tissues removed from the site showed gap junctional formation between myocyte and hMSC elements [11]. Finally, the stem cells stained positively for vimentin, indicating that they were mesenchymal and positively for human CD44 antigen, indicating that they were of human origin (Fig. 12) [11].

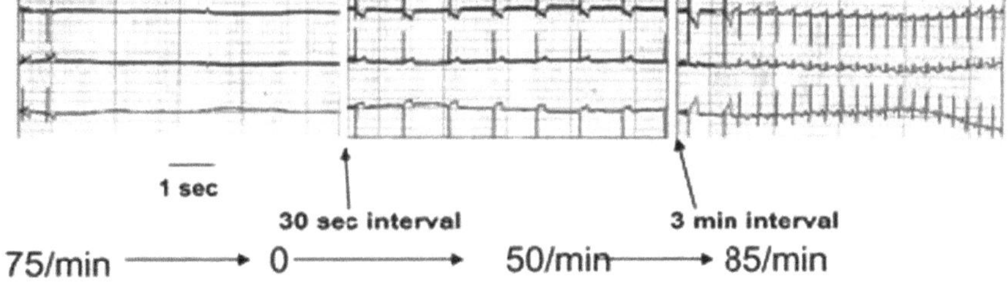

Fig. 11. Human mesenchymal stem cells (hMSC)-based pacemaker function in canine heart in situ. *Top to bottom* ECG leads I, II, III, AVR, AVL, and AVF. *Left* Last two beats in sinus rhythm and onset of vagal stimulation (*arrow*) causing sinus arrest in a dog studied 7 days after implanting mHCN2- transfected hMSCs in LV anterior wall epicardium. ***Middle*** During continued vagal stimulation, an idioventricular escape focus emerges, having a regular rhythm. *Right* On cessation of vagal stimulation (***arrow***), there is a postvagal sinus tachycardia (reprinted by permission from reference [8]).

In a preliminary study [7], we have followed the function of hMSC-based biological pacemaking through 6 weeks post-implantation and have found that the rate generated is stable. Equally importantly, we have used staining for immune globulin and for canine lymphocytes to determine if rejection of the hMSCs were occurring. At 2 week and 6 week time points, there was no evidence for humoral or cellular rejection. This is consistent with the earlier work of Liechty et al. [4] suggesting that hMSCs may be immunoprivileged. If more detailed investigation demonstrates this to be the case, then it would abrogate any need for immunosuppression. Certainly any need for immunosuppressive drugs would be a major detriment to cell therapy approaches, and would argue strongly in favor of staying with electronic pacemakers.

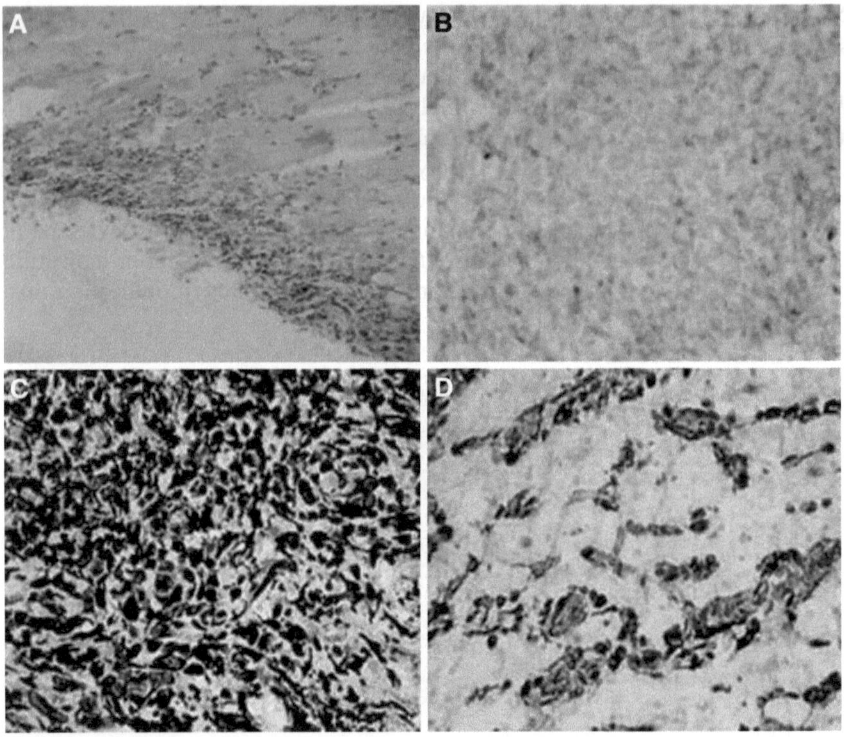

Fig. 12. **a** H&E stain showing basophilicstained stem cells and normal myocardium. **b** and **c** show, respectively, vimentin and CD44 staining of a node of human mesenchymal stem cells in canine myocardium. **d** Detail of vimentinstained cells interspersed with myocardium. Magnification 100·(**a**) and 400·(**b–d**) (reprinted by permission from reference [8]).

4 Conclusions

Much has been accomplished in showing that both viral and hMSC platform approaches are effective in generating biological pacemakers. But as we stated at the outset, much remains to be done. We have summarized this as follows: [3, 12, 13]. Whether the approach is with virus or stem cell, we need evidence that it is/is not superior to the electronic pacemaker in terms of adaptability to the body's physiology and duration of effectiveness. We also need evidence regarding long-term incidence of inflammation, infection, rejection, and neoplasia, and for or against long-term proarrhythmic potential. We need to understand if the construct is localized at the site of implantation or migrates to other sites. Other toxicities of which we may not be aware need to be looked for and delivery systems must be optimized. In addition, for hMSCs (and indeed for any stem cells) we need evidence regarding persistence of the administered cell types versus their differentiation into other cell types. In the latter event, evidence regarding persistence of pace-maker function, in terms of physiologic expression, current generated, and coupling to adjacent cells, will be essential.

Despite the challenges remaining, one point should be clear at present: that is, I_f alone is adequate—whether administered via virus or via plat-form—to drive the heart. Given this information, and given the need to use the simplest possible system to generate pacemaker function, we plan to persist with this approach. We do this because of the belief that the more complexity that is brought into the system, the more that can go wrong. Hence, we wish to avoid additional potentially confounding components unless further research proves them absolutely essential.

Acknowledgments. The authors express their gratitude to Ms Laureen Pagan for her careful attention to the preparation of the manuscript. The studies described were supported by USPHS-NHLBI grants HL 28958 and HL 67101, and by Guidant Corporation.

References

1. Biel M, Schneider A, Wahl C (2002) Cardiac HCN channels: structure, function, and modulation. Trends Cardiovasc Med 12:206–212
2. Chen J, Mitcheson JS, Tristani-Firouzi M, Lin M, Sanguinetti MC (2001) The S4–S5 linker couples voltage sensing and activation of pacemaker channels. Proc Natl Acad Sci USA 98:11277–11282
3. Cohen IS, Brink PR, Robinson RB, Rosen MR (2005) The why, what, how and when of biological pacemakers. Nat Clin Pract Cardiovasc Med 2:374–375
4. Liechty KW, MacKenzie TC, Shaaban AF, Radu A, Moseley AM, Deans R, Marshak DR, Flake AW (2002) Human mesenchymal stem cells engraft and demonstrate site specific differentiation after in utero implantation in sheep. Nat Med 6:1282–1286
5. Plotnikov AN, Sosunov EA, Qu J, Shlapakova IN, Anyukhovsky EP, Liu L, Janse MJ, Brink PR, Cohen IS, Robinson RB, Danilo P Jr, Rosen MR (2004) A biological pacemaker implanted in the canine left bundle branch provides ventricular escape rhythms having physiologically acceptable rates. Circulation 109:506–512
6. Plotnikov AN, Shlapakova IN, Kryukova Y, Bucchi A, Pan Z, Danilo P Jr, Brink PR, Cohen IS, Robinson RB, Rosen MR (2005a) Comparison of mHCN2 and mHCN2-E324A genes as biological pacemakers. Circulation 112:II-126 (Abstract)
7. Plotnikov AN, Shlapakova IN, Szabolcs MJ, Danilo P Jr, Lu Z, Potapova I, Lorell BH, Brink PR, Robinson RB, Cohen IS, Rosen MR (2005b) Adult human mesenchymal stem cells carrying HCN2 gene perform biological pacemaker function with no overt rejection for 6 weeks in canine heart. Circulation 112:II-221 (Abstract)

8. Potapova I, Plotnikov A, Lu Z, Danilo P Jr, Valiunas V, Qu J, Doronin S, Zuckerman J, Shlapakova IN, Gao J, Pan Z, Herron AJ, Robinson RB, Brink PR, Rosen MR, Cohen IS (2004) Human mesenchymal stem cell as a gene delivery system to create cardiac pacemakers. Circ Res 94:952–959
9. Qu J, Barbuti A, Protas L, Santoro B, Cohen IS, Robinson RB (2001) HCN2 overexpression in newborn and adult ventricular myocytes: distinct effects on gating and excitability. Circ Res 89:e8–e14
10. Qu J, Plotnikov AN, Danilo P Jr, Shlapakova I, Cohen IS, Robinson RB, Rosen MR (2003) Expression and function of a biological pacemaker in canine heart. Circulation 107:1106–1109
11. Qu J, Kryukova Y, Potapova IA, Doronin SV, Larsen M, Krishnamurthy G, Cohen IS, Robinson RB (2004) MiRP1 modulates HCN2 channel expression and gating in cardiac myocytes. J Biol Chem 279:43497–43502
12. Rosen M (2005) Biological pacemaking: inour lifetime? Heart Rhythm 2:418–428
13. Rosen MR, Brink PR, Cohen IS, Robinson RB (2004) Genes, stem cells and biological pacemakers. Cardiovasc Res 64:12–23
14. Valiunas V, Doronin S, Valiuniene L, Potapova I, Zuckerman J, Walcott B, Robinson RB, Rosen MR, Brink PR, Cohen IS (2004) Human mesenchymal stem cells make cardiac connexins and form functional gap junctions. J Physiol 555:617–626

Gene Therapy to Create Biological Pacemakers

Gerard J.J. Boink[1,2], Jurgen Seppen[3], Jacques M.T. de Bakker[1,2], and Hanno L. Tan[1]

[1] Department of Clinical and Experimental Cardiology, Academic Medical Center, University of Amsterdam, Meibergdreef 9, 1105 AZ Amsterdam, The Netherlands
h.l.tan@amc.uva.nl
[2] Interuniversity Cardiology Institute Netherlands, Utrecht, The Netherlands
[3] Liver Center, Academic Medical Center, University of Amsterdam, Amsterdam, The Netherlands

Abstract. Old age and a variety of cardiovascular disorders may disrupt normal sinus node function. Currently, this is successfully treated with electronic pacemakers, which, however, leave room for improvement. During the past decade, different strategies to initiate pacemaker function by gene therapy were developed. In the search for a biological pacemaker, various approaches were explored, including β_2-adrenergic receptor overexpression, down regulation of the inward rectifier current, and overexpression of the pacemaker current. The most recent advances include overexpression of bioengineered ion channels and genetically modified stem cells. This review considers the strengths and the weaknesses of the different approaches and discusses some of the different viral vectors currently used.

Keywords: Cardiac arrhythmia therapy, Ion channels, Pacemakers, Gene therapy, Sinoatrial node.

1 Introduction

Electronic pacemakers are of great value in the therapy of cardiac conduction disease. These devices have become more and more sophisticated over the past years, but there are shortcomings. Items that need improvement include the lack of autonomic modulation of the heart rate, the limited battery life, unstable electrode position, and electronic or magnetic interference. A biological pacemaker may circumvent these adverse effects and would be an ideal alternative. Given the good, although suboptimal, performance of the electronic pacemaker, quality standards for the bio-engineered version must be high [40].

2 Autonomic Activity of the SA Node

A proper understanding of the physiological pacemaker, the sinoatrial (SA) node, is of great importance when different approaches in the creation of a biopacemaker are considered. The SA node is a heterogeneous structure composed of specialized cardiomyocytes and a high level of connective tissue. The activity in this node is driven by a spontaneous change in the membrane potential, called the slow diastolic

J.A.E. Spaan et al. (Eds.): Biopacemaking, BIOMED, pp. 79–93, 2007.
springerlink.com

depolarization or phase 4 depolarization. This phase 4 depolarization results in the formation of action potentials, thereby triggering the contraction of the heart (Fig. 1a). The most important current underlying this process is the "funny current" or I_f. A family of hyperpolarization-activated cyclic nucleotide-gated (HCN) channels is believed to underlie this inward current. There are four HCN isoforms which are all expressed in the heart, but expression levels vary among regions [30, 41]. In the rabbit SA node, HCN4 is the dominant transcript representing more than 81% of total HCN

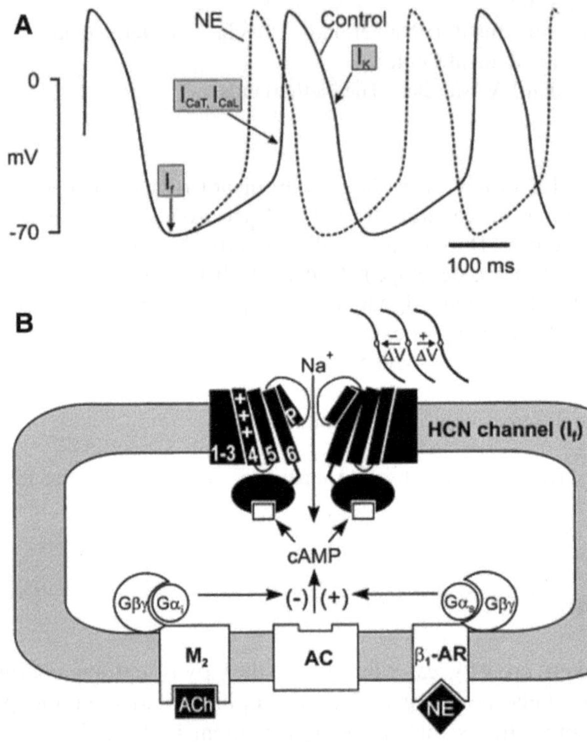

Fig. 1. **a** Potential changes in the SA node and some ion channels that are involved. The process of slow diastolic depolarization is initiated by I_f, which is activated upon hyperpolarization. HCN channels are the main proteins underlying this current. When the threshold is reached, the action potential starts as a result of the opening of T-type and L-type calcium channels. Repolarization occurs mainly due to the opening of K^+ currents. A faster rate is shown by the effect of β-adrenergic stimulation with norepinephrine (NE), and results from the increase in the slope of phase 4 depolarization. **b** Regulation of HCN channel activity by alterations of intracellular cyclic adenosine monophosphate (cAMP). β_1-Adrenergic receptor (β_1-AR) stimulation increases cAMP levels, as a result of G-protein coupled regulation of adenylyl cyclase (AC) activity. M_2-muscarinic receptor stimulation results in the opposite. Cyclic AMP binds to HCN channels near the amino terminus, where it accelerates activation kinetics and shifts the voltage dependence of activation to more positive voltages (as shown on top, see also Fig. 3) (from reference [2], with permission).

mRNA [42]. The activity of HCN channels is controlled by the cyclic adenosine monophophate (cAMP)-binding site which allows alteration of activation kinetics by β-adrenergic and muscarinic stimulation (Fig. 1b). By this mechanism, channel activity might be increased or decreased. This plays an important role in the autonomic regulation of heart rate [7]. However, I_f is not the only current contributing to the pacemaker cell membrane potential. Other inward and outward currents are involved as well. Any increase in inward and/or decrease in outward current may initiate or accelerate the process of phase 4 depolarization [2].

3 How Can a Biological Pacemaker Be Created?

To induce the spontaneous release of action potentials in normal cardiac cells three main gene therapy strategies have been developed: (1) upregulation of β_2-adrenergic receptors (β_2-AR) [11]; (2) knock-down of outward potassium current (I_{K1}) [26]; (3) overexpression of inward cation current (If) [35]. In addition to these strategies, we will discuss two cell therapy approaches.

3.1 Upregulation of β_2-Adrenergic Receptors

The proof of concept for the creation of a biopacemaker by modulating chronotropy was provided in 2001 by Edelberg et al [11]. They showed for the first time increased contraction rates in murine cardiac myocytes after introduction of a plasmid with the human β_2-AR gene into these cells [10]. Later in vivo experiments in mouse and swine were conducted in which injections of the β_2-AR carrying construct into the right atrium increased heart rates by ~40 and ~50%, respectively [10, 11]. Although these experiments demonstrated that gene therapy is able to alter cardiac rhythm in intact hearts of large animals, this approach was designed as a proof-of-principle and, accordingly, lacked a design appropriate for clinical applicability. First, only transient expression is induced by the use of these delivery platforms. Second, modulation of β-adrenergic responsiveness will only modulate the rate at which the native pacemaker system fires. In case of a diseased SA node, this altered responsiveness might worsen the situation, resulting in additional arrhythmias. Patients with sick sinus syndrome, now treated with electronic pacemakers, have a disrupted sinus node function, resulting in disease causing bradycardias (slow heart rates) in conjunction with atrial tachycardias (fast heart rates). This condition cannot be treated with upregulation of β_2-adrenergic receptors, since this may worsen the tachycardias.

3.2 Knock Down of Outward, Hyperpolarizing Current (I_{K1})

There are two strategies by which the resting membrane potential (RMP) may be disturbed to generate spontaneous slow diastolic depolarization. One is the knock-down of the hyperpolarizing inward rectifier potassium channels. These channels are abundantly expressed in the working myocardium of the atrium and ventricle but not in the SA node and they play an important role in repolarization and stabilization of the RMP. A knock down of these channels results in depolarization of the RMP, which liberates endogenous pacemaker activity [27].

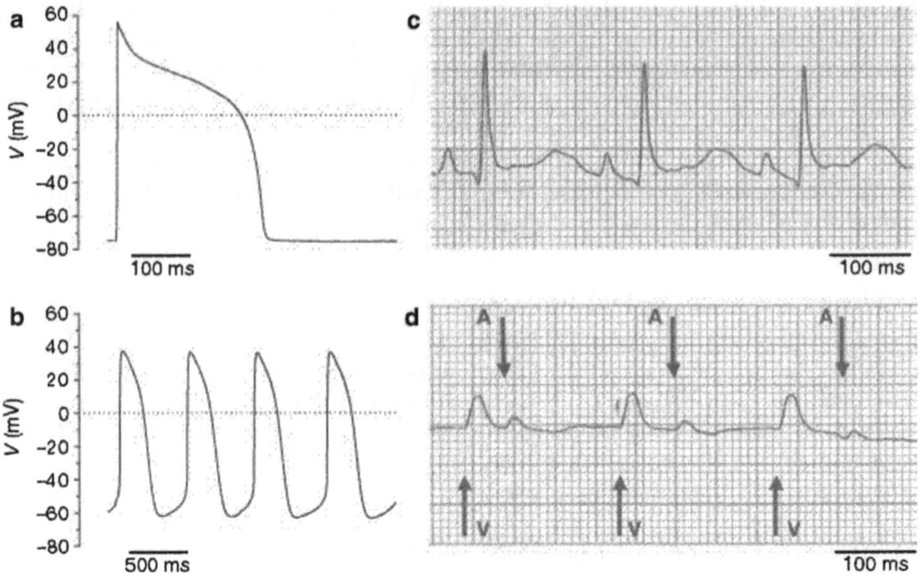

Fig. 2. Suppression of Kir2.1 channels liberates pacemaker activity. **a** Stable action potentials initiated by external depolarizing stimuli in control ventricular myocytes. **b** Spontaneous action potentials with a SA nodal phenotype in Kir2.1AAA transduced myocytes. **c** Control electrocardiogram of normal sinus rhythm. **d** Electrocardiogram of ventricular rhythms 72 h after transduction with Kir2.1AAA. P waves (*arrow A*) and wide QRS complexes (*arrow V*) are uncoupled, demonstrating that ventricular activation does not proceed along the normal conduction pathway (from reference [26], with permission).

The first in vivo proof-of-concept of this approach was provided in 2002 by Miake et al. [26], who built a dominant negative construct by replacing three amino acid residues in the pore of Kir2.1. Four Kir2.1 subunits normally assemble to form tetrameric inward rectifier potassium channels. Adenoviral (Ad) vectors were used to target the left ventricular cavity of guinea pigs. After 3–4 days, in vivo Kir2.1AAA transduced myocytes were isolated and a reduction of about 80% of I_{K1} was found. Successfully transduced cells increased their pacing rate in response to β-adrenergic stimulation, and in 40% of the animals, premature ventricular beats occurred. Figure 2 b illustrates the development of spontaneous activity in a ventricular myocyte after reducing I_{K1}. The action potential of an untreated myocyte is depicted in panel a. Panel d shows the ensuing rhythm in the intact animal after treatment, while panel c is the ECG of a control animal. One major concern regarding this strategy is that reduction in repolarizing currents may result in excess prolongation of repolarization which constitutes a proarrhythmic effect, potentially causing torsades de pointes ventricular tachyarrhythmias that could degenerate to ventricular fibrillation [46]. Loss-of-function mutations in Kir2.1 are clinically manifested in Andersen–Tawil syndrome (also known as Long-QT Syndrome type 7). This syndrome consists of mild QT interval prolongation, prominent ECG U waves, frequent ventricular ectopy and polymorphic ventricular tachycardia, in conjunction with extracardiac features such as periodic paralysis [52, 56]. Another major concern surrounding I_{K1} knock-down is, that theoretical considerations predict that impaired repolarization of resting membrane potential following from reduction in I_{K1} may result in membrane potential oscillations that may culminate in cardiac arrhythmias.

3.3 Overexpression of Inward Depolarizing current (I_f)

The other approach to depolarize the RMP is the introduction of I_f. It is believed that I_f is the primary pacemaker current in the SA node, where it is mainly generated by HCN4, and, to a smaller extent, HCN2 and HCN1. The first effects of HCN overexpression were reported in 2001 by Qu et al. [35], who showed an increased beating frequency in neonatal ventricular myocytes after transduction with Ad-mHCN2. They also demonstrated that the potential for autonomic modulation is retained. In an I_f activation curve, positive effects of cAMP were clearly demonstrated by a stronger activation of the current at the same membrane potential (Fig. 3). After this in vitro proof of principle, the same vectors were used for in vivo experiments. In four dogs, the left atrium (LA) was injected with Ad-mHCN2 and spontaneous LA rhythms were recorded after vagal stimulation-induced sinus arrest (Fig. 4) [36]. The same vectors were injected into the left bundle branch of seven dogs. During vagal stimulation, all of the HCN2 treated dogs demonstrated more rapid escape rhythms that originated from the left ventricle, compared to slower controls [33].

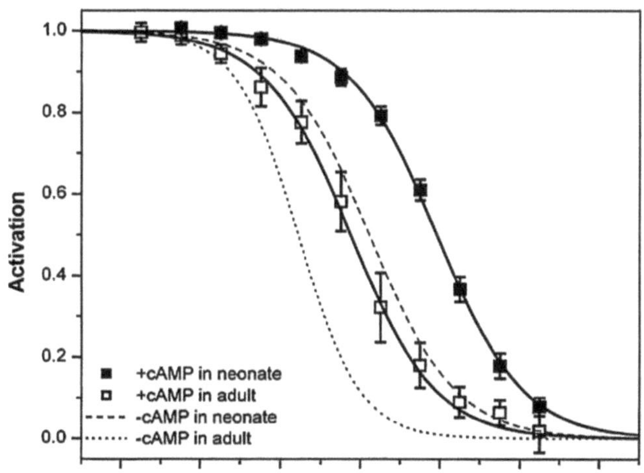

Fig. 3. Rightward shift of the I_{HCN2} activation curve in neonatal and adult rat cardiac myocytes due to additional intracellular cAMP. This graph shows that, at a certain potential, channel activity is increased as a result of higher cAMP levels (from reference [35], with permission).

Clearly, an uncontrolled increase in heart rate, by whichever gene therapy strategy, may cause deleterious effects on cardiovascular function. This has been made particularly obvious by recent clinical trials of the prototype I_f blocker ivabradine. In these trials, ivabradine exhibited clinical benefits when used with the aim of preventing angina pectoris through heart rate reduction. Conversely, it may be envisaged that the availability of ivabradine may be exploited to fine tune the pulse rate when used in conjunction with HCN-based gene therapy for the creation of a biopacemaker [3, 7].

3.4 Cell Therapy

In addition to these gene therapy strategies, various cell therapy approaches have yielded remarkable results. Here, we discuss some of these approaches as well as the genetic modification of applied stem cells.

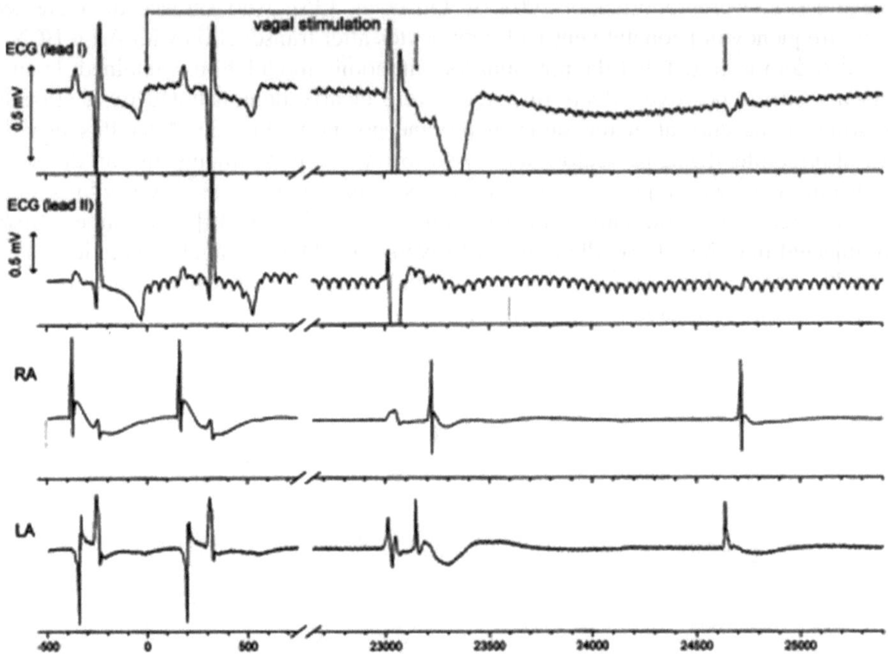

Fig. 4. Electrocardiograms of leads I, II, RA and LA (overlying injection site). The initial two beats represent normal sinus rhythm in an anesthetized dog previously injected in the LA with Ad-HCN2-GFP. ECGs are interrupted during vagal stimulation (time-point 0). Shortly therafter, asystole was induced. An idioventricular escape beat and two ectopic P waves (*arrows*, lead I) originating from the LA are shown in the last part of the ECGs. Electrical activity in LA precedes that in RA (from reference [36], with permission).

Adult human mesenchymal stem cells (hMSCs) are multipotent, which means that they are, in contrast to embryonic stem cells, only able to differentiate into mesenchymally derived cell lineages. Therefore, these cells always have to be genetically modified when used to generate a biopacemaker. Potapova et al. [34] described the delivery of HCN2 into hMSCs by electroporation. Unfortunately, electroporation induces only transient expression of the HCN channels. To induce long-term gene expression in these cells, a different method of genetic modification is required. However, hMSCs expressing HCN2 were injected subepicardially into the canine left ventricular wall and faster escape rhythms compared to controls, were recorded during sinus arrest. One should realize that hMSCs are not equipped with the complete set of ion channels to either initiate an AP or to hyperpolarize the RMP to a membrane potential range where HCN channels are active. Therefore, electric coupling to cardiac myocytes is of crucial importance. When the cardiomyocytes are well coupled to stem cells, they are able to hyperpolarize the hMSCs electrotonically, and activate HCN channel opening. This could be followed by a slow phase 4 depolarization in the hMSCs, which would result in the initiation of APs in cardiac myocytes (Fig. 5).

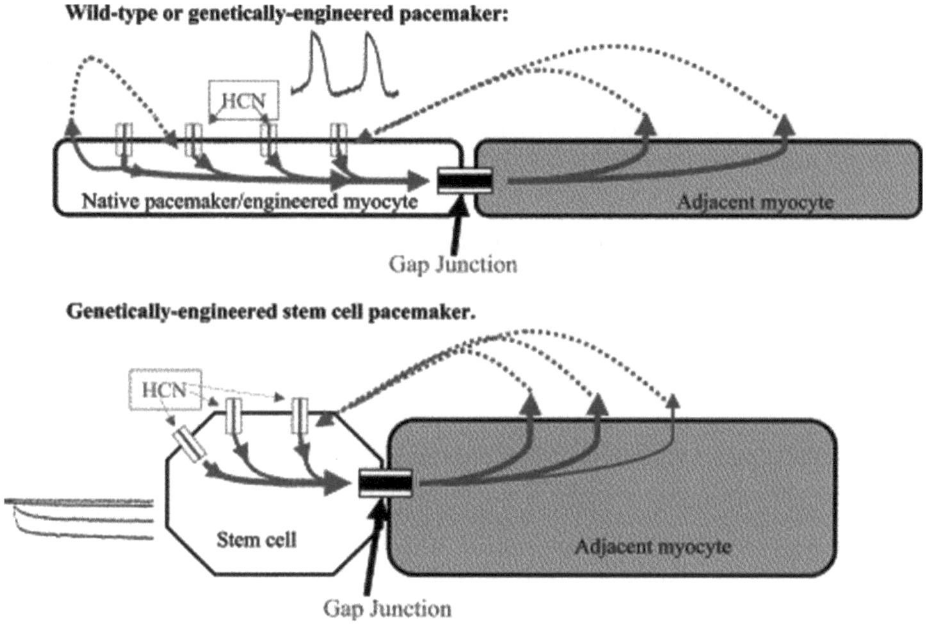

Fig. 5. Pacemaker activity initiated from myocyte or human mesenchymal stem cell (*hMSC*). Top Sinoatrial node cell or gene therapy targeted myocyte connected via gap junctions to a surrounding myocyte. Action potentials (*inset*) are initiated by slow diastolic depolarization resulting from current flowing through HCN channels. *Bottom* HCN channels overexpressed in hMSCs, these channels can only be activated if the membrane potential is hyperpolarized by the adjacent myocyte. Channel activation will result in excitation of the adjacent myocyte via gap junctions, which will initiate action potential formation (from reference [39], with permission).

An advantage of hMSCs, is that they are possibly immunoprivileged, i.e., they have not elicited major immune responses in limited studies [25]. However, a drawback is that it is uncertain how these cells may differentiate over time after transplantation. It is possible that these cells differentiate into cardiac cells but differentiation into other cells cannot be ruled out [37]. In addition, concerns are rising concerning the risk of neoplasia, rejection, or migration to other sites [39].

Embryonic stem cells (ESCs) are pluripotent, allowing them to differentiate into virtually any cell type. Therefore, genetic engineering of these cells to create a biological pacemaker is not always necessary, as differentiation can be directed by changing culturing conditions [20]. However, genetic modification of hESCs provides a great opportunity in the selection of an appropriate subpopulation. Using the α-myosin heavy chain (α-MHC) promoter driving the expression of the enhanced green fluorescent protein (EGFP), pacemaker-like cardiomyocytes could be selected based on fluorescence intensities. In these experiments, a quantitative relation between EGFP expression and atrial-and pacemaker-like phenotypes of the hEBSc was shown. Ventricular-like cells proved to be exclusively EGFP negative [23]. In addition to this application, the combination of the α-MHC promoter, the EGFP cassette and a second transgene (e.g., a HCN gene) provides possibilities for both tracing the optimal subpopulation and further fine-tuning of the hESC properties. Nevertheless, much has been written about the socio-political fear about the use of these cell and more technical concerns regarding the expected requirement for additional immunosuppressive treatment [38, 43].

4 Outlines for a Biological Pacemaker

If a biological pacemaker is to compete with current therapy, various requirements and safety issues have to be fulfilled. First, there is a need for autonomic regulation. This may be accomplished by cardiac gene therapy. If one of the HCN channels is selected, autonomic modulation will occur by adrenergic or muscarinic receptor pathways that are available in every cardiac myocyte. Changes in intracellular cAMP caused by these pathways will then alter channel activity.

Second, the site of implantation is important. In electronic pacemakers, implantation sites are restricted to areas where stable lead positions can be obtained. With the gene and cell-therapy approaches, it is anticipated that there is much more freedom to choose a suitable position. This is an advantage if there is cardiac comorbidity, and arrhythmogenic substrates are present. In these patients, the best avenue with minimal arrhythmic potential could be selected via catheter-based intra-cardiac mapping.

Two other issues are important: duration of effect and bio-safety. The functional duration of pacemaking should be comparable to current (and future) life spans of the batteries that are used in electronic pacemakers. In a gene therapy approach, this requires stable and long-term expression of the transgene. When stem cells are used, the survival, migration and dedifferentiation of these cells are of additional importance. Gene therapy could provide solutions addressing these problems. For example, in ischemic hearts, hMSCs survival has improved by transfecting these cells with a hypoxia-regulated heme oxygenase-1 (HO-1) plasmid. Heme oxygenase-1 is a key component inhibiting inflammatory cytokines and proapoptotic factors which are commonly liberated during hypoxia and reoxygenation [47].

With regards to bio-safety, a minimal risk for infections and neoplasias should be guaranteed. The selection of an appropriate vector system is importantly determined by these safety requirements. In summary, the ideal system combines stable long-term expression with low immunogenicity and zero carcinogenicity.

Table 1. Different viral vectors to target cardiac tissue

Vector	Viral genome	Cloning capacity	Inflammation	Vector genome forms	Main limitations	Main advantages
Lentivirus	RNA	8.0 kb	Low	Integrated	Fragile particle; insertional mutagenesis	Long-term expression. Broad cell tropism. Noninflammatory and nonpathogenic
AAV	ssDNA	<5.0 kb	Low	Episomal (<90%), Integrated (<10%)	Small cloning capacity; preexisting immunity	Stable particle. Broad cell tropism; noninflamatory and nonpathogenic
Adenovirus	dsDNA	7.9 kb	High	Episomal	Capsid mediates an inflammatory response; preexisting immunity	Highly efficient transduction; large cloning capacity; high titer

5 Delivery Platforms to Target Cardiac Tissue

Adenoviral vectors are often used in cardiac gene therapy. They can be produced with high titers and transduce cardiac myocytes efficiently [4, 22]. The transient nature of adenoviral induced gene expression and the expected inflammatory response to adenoviral gene products renders them unsuitable for the creation of a biological pacemaker [19, 51]. The development of helper dependent Ad vectors might circumvent some of these problems. In these vectors, no adenoviral genes are present and they can mediate therapeutic gene expression over much longer periods than earlier generation Ad vectors. However, because helper dependent Ad vectors remain episomal, it is uncertain whether they will mediate the long-term gene expression needed for the generation of a stable biopacemaker [53] (Table 1).

Vectors based on adeno-associated viruses (AAV) can mediate long-term expression of therapeutic transgenes. There is no transfer of AAV genes and immunogenicity is therefore limited. AAV vectors have been demonstrated to mediate efficient transduction of cardiac myocytes [45]. However, AAV applicability is limited by their relatively small packaging capacity (less than 5 kb) [9]. As with adenoviral vectors, AAV vectors remain episomal which makes it uncertain whether they will be able to mediate gene expression over very long periods.

In our view, for a biopacemaker to compete with electronic pacemakers, integration of therapeutic genes in the host cell genome is necessary because this will ensure long-term function. Gene therapy vectors based on retroviruses are able to integrate into the host cell genome and mediate long-term expression. Lentiviral vectors are retroviral vectors based on human immunodeficiency virus. These vectors have been shown to mediate efficient gene transfer in cardiomyocytes in vitro and in vivo [13, 57]. However, when lentiviral vectors are put in a clinical perspective, biosafety is a major issue. Multiple safety steps have been built into the development of these vectors. As with AAV vectors, no viral genes are transferred. The risk of generation of recombinant replication competent virus is practically absent, because all vector components are encoded on separate plasmids. Furthermore, the transfer vector contains large deletions in the viral long terminal repeats which minimizes the chance of possible interaction between the integrated vector and wild type HIV [31, 58]. The most important fear, however, is of insertional mutagenesis. Although tumor formation had never been observed in rodents treated with retroviral vectors, three patients with X-linked, severe combined immune deficiency (X-SCID), that were treated using retroviral gene transfer, developed T-cell leukemia. However, the combination of vector insertion near the oncogene Lmo2 and a transgene (gamma common chain) that is involved in cell development, were probably required for tumor formation [15]. Furthermore, in this trial, few gene corrected cells had to undergo many cell divisions to repopulate the entire immune system of the patients.

Because biopacemaker gene therapy does not require cell division of transduced cells and because the therapeutic genes are not involved in cell proliferation, the risk of tumor formation seems low in this case [6]. Lentivectors are therefore generally believed to be a safe delivery platform. This is further supported by the fact that insertional mutagenesis never has been identified in HIV positive patients. In addition, heart tissue is composed of non-dividing cells, making them relatively resistant to oncogenesis.

The use of lentiviral vectors in the ex vivo generation of biopacemaker cells, such as the use of hMSCs or hESCs should also be considered. Both cell types are efficiently transduced by these vectors and, in hESCs, stable genetic modification was even combined with a preserved cardiogenic potential. In these experiments, a CAG promoter was used to circumvent problems of gene silencing in these cells [24, 50, 54].

6 Vector Delivery Techniques

Currently available vector delivery techniques to target a focal region include direct intramyocardial injection, subepicardial injection and epicardial administration [8]. Intramyocardial injection of viral vectors results in highly localized transgene activity. When Ad vectors were injected in the apex of the left ventricle, transgene expression was limited to a few millimeters surrounding the injection site [18]. Other injection sites have also been proven to be reachable. Ad vectors were injected into the left bundle branch [33] and into the left atrium (subepicardially) [36]. Although major complications were not mentioned, site specificity of these approaches was not analyzed here. Atrial injections might be most challenging, but particularly relevant, since this is the site of the natural pacemaker. The thinness of the atrial wall, as experienced during sub-epicardial plasmid injections in the right atrium of pigs, showed both the difficulty and the feasibility of this approach [12].

A highly innovative epicardial administration of Ad vectors appeared perfectly suited for atrial gene transfer. A gelatinous poloxamer matrix complexed to the virus was painted on both atria after they were exposed by a median sternotomy followed by a pericardium incision. A 100% transmural gene transfer was obtained in all atrial regions, using dilute trypsin concentrations combined with the poloxamer matrix. This elegant approach showed no evidence of reporter gene expression in the cardiac ventricles, lungs, liver, spleen, kidney, gonads or skeletal muscle [21]. More specific targeting of epicardially applied viral vectors may hold the greatest promise. This could be reached with engineered viral vectors that specifically bind to receptors present on atrial cardiomyocytes or by the use of magnetic carriers in combination with externally applied magnetic fields [55] (this approach has similarities with recently introduced methods of intracardiac catheter delivery using magnetic fields [48]).

7 Which Genes Should Be Delivered to Create a Biological Pacemaker?

Classical pharmacology is not only about the active substance, but also about the formulation as a whole. This is not different in gene therapy. It is necessary to select the most appropriate viral vector system produced at the best quality as discussed above, but, within this vector, the specificity of transcription is of equal importance. Instead of a constitutive promoter which has been used in most of the previous gene therapy approaches, conditional gene expression may be preferred. Cardiac specific promoters will be useful for cell-type-specific expression [16]. Some cardiac promoters are well described, e.g., the myosin light chain 2v (MLC-2v) promoter (ventricular specific) [14, 32, 45], the α-myosin heavy chain promoter (atrial and ventricular specific) [14], and the atrial natriuretic factor (ANF) promoter (atrial specific) [49]. However, at present, constitutive promoters are suitable for experimental purposes.

About the active substance, the therapeutical gene, we can be more specific. In our opinion, HCN channels are the most promising candidates to deliver. This is because they incorporate autonomic modulation by direct allosteric interaction with cAMP and they are only activated upon hyperpolarization. The latter results in channel inactivity during the action potential plateau phase. The knock-down of the inward rectifier potassium current, in contrast, is known to prolong action potentials and may cause excess prolongation of repolarization resulting in torsade de pointes, while oscillations in resting membrane potential may also constitute inappropriate membrane depolarizations and cardiac arrhythmias.

If the delivery of a HCN channel is considered, HCN2 and HCN4 are primary candidates. HCN4 is highly expressed in the SA-node and a mutation in this channel is associated with familial sinus bradycardia [29]. Activation of HCN2, on the other hand, may have the advantage of being much faster (i.e., having a shorter response time) [44]. Single channel analysis has shown that I_f in human atrial myocytes closely resemble characteristics of HCN4 or HCN2 + HCN4 [28].

A bio-engineered designer channel could be the ultimate transgene [1] and some interesting results in this research area are now becoming available. The group of Marbán described the conversion of a depolarization-activated K-channel into a hyperpolarization-activated channel, which also became permeable to Na. With multiple mutations, this yielded a mutant channel with a pacemaker-current phenotype [17]. The group of Rosen described improved pacemaker function using the E324A point mutation in the S4–S5 linker (linker of the fourth and fifth transmembrane segments) of murine HCN2 (mHCN2). This mHCN2–E324A channel induces faster and more positive pacemaker current activation compared to normal mHCN2. This bioengineered channel was already tested in vivo, when it was applied in the left bundle branch of four dogs using Ad vectors. Treated dogs showed faster rhythms with stronger catecholamine sensitivity than mHCN2 and controls [5]. If these mutations eventually lead to precise rate modulation, issues on the regulation of gene expression might be irrelevant.

8 Conclusions

For more than a decade, most of the cardiac gene therapy research has focused on Ad vectors. If a bio-logical pacemaker is to challenge the current standard of an electronic pacemaker, a long-term expression system is needed. For this purpose, lentiviral vectors may be the most suitable, but further research is needed to evaluate if stable, long-term, pacemaker function can be initiated by HCN-gene delivery from these vectors. The electric coupling between gene therapy targeted and surrounding cells, is another major research area that needs to be explored. Human mesenchymal or embryonic stem cells could also provide a useful tool here to alter and control this electrical coupling.

Bioengineered HCN channels have great potential. Although these channels are still in an early stage of development, they may allow fine-tuning of the heart rate range, making the clinical perspective of biological pacemakers increasingly realistic.

Acknowledgments. This work was supported by the Netherlands Heart Foundation (NHS 2005B180 to HLT and JMTdB), the Royal Netherlands Academy of Arts and Sciences KNAW (HLT), and the Bekales Foundation (HLT).

References

1. Azene EM, Xue T, Marbán E, Tomaselli GF, Li RA (2005) Non-equilibrium behavior of HCN channels: insights into the role of HCN channels in native and engineered pacemakers. Cardiovasc Res 67:263–273
2. Biel M, Schneider A, Wahl C (2002) Cardiac HCN channels: structure, function, and modulation. Trends Cardiovasc Med 12:206–212
3. Borer JS (2004) Drug insight: if inhibitors as specific heartrate-reducing agents. Nat Clin Pract Cardiovasc Med 1: 103–109
4. Brenner M (1999) Gene transfer by adenovectors. Blood 94:3965–3967
5. Bucchi A, Plotnikov AN, Shlapakova I, Danilo P Jr., Kryukova Y, Qu J, Lu Z, Liu H, Pan Z, Potapova I, Ken-knight B, Girouard S, Cohen IS, Brink PR, Robinson RB, Rosen MR (2006) Wild-type and mutant HCN channels in a tandem biological-electronic cardiac pacemaker. Circulation 114:992–999
6. Dave UP, Jenkins NA, Copeland NG (2004) Gene therapy insertional mutagenesis insights. Science 303:333
7. DiFrancesco D (2005) Cardiac pacemaker I(f) current and its inhibition by heart rate-reducing agents. Curr Med Res Opin 21:1115–1122
8. Donahue JK, Kikuchi K, Sasano T (2005) Gene therapy for cardiac arrhythmias. Trends Cardiovasc Med 15:219–224
9. Duan D, Yue Y, Engelhardt JF (2001) Expanding AAV packaging capacity with trans-splicing or overlapping vec-tors: a quantitative comparison. Mol Ther 4:383–391
10. Edelberg JM, Aird WC, Rosenberg RD (1998) Enhancement of murine cardiac chronotropy by the molecular transfer of the human beta2 adrenergic receptor cDNA. J Clin Invest 101:337–343
11. Edelberg JM, Huang DT, Josephson ME, Rosenberg RD (2001) Molecular enhancement of porcine cardiac chronotropy. Heart 86:559–562

12. Fishbein I, Stachelek SJ, Connolly JM, Wilensky RL, Alferiev I, Levy RJ (2005) Site specific gene delivery in the cardiovascular system. J Control Release 109:37–48
13. Fleury S, Simeoni E, Zuppinger C, Déglon N, von Segesser LK, Kappenberger L, Vassalli G (2003) Multiply attenuated, self-inactivating lentiviral vectors efficiently deliver and ex-press genes for extended periods of time in adult rat cardiomyocytes in vivo. Circulation 107: 2375–2382
14. Franz WM, Rothmann T, Frey N, Katus HA (1997) Analysis of tissuespecific gene delivery by recombinant adenoviruses containing cardiacspecific promoters. Cardiovasc Res 35:560–566
15. Hacein-Bey-Abina S, von Kalle C, Schmidt M, Le Deist F, Wulffraat N, McIntyre E, Radford I, Villeval JL, Fraser CC, Cavazzana-Calvo M, Fischer A (2003) A serious adverse event after successful gene therapy for X-linked severe combined immunodeficiency. N Engl J Med 348:255–256
16. Heine HL, Leong HS, Rossi FM, McManus BM, Podor TJ (2005) Strategies of conditional gene expression in myocardium: an overview. Methods Mol Med 112:109–154
17. KashiwakuraY, ChoHC, AzeneE, MarbánE(2005)Creation of a synthetic pacemaker channel. Circulation 112:U147
18. Kass-Eisler A, Falck-Pedersen E, Alvira M, Rivera J, Buttrick PM, Wittenberg BA, Cipriani L, Leinwand LA (1993) Quantitative determination of adenovirus-mediated gene delivery to rat cardiac myocytes in vitro and in vivo. Proc Natl Acad Sci USA 90:11498–11502
19. Kass-Eisler A, Falck-Pedersen E, Alvira M, Rivera J, Buttrick PM, Wittenberg BA, Cipriani L, Leinwand LA (1993) Quantitative determination of adenovirus-mediated gene delivery to rat cardiac myocytes in vitro and in vivo. Proc Natl Acad Sci USA 90:11498–11502
20. Kehat I, Khimovich L, Caspi O, Gepstein A, Shofti R, Arbel G, Huber I, Satin J, Itskovitz-Eldor J, Gepstein L (2004) Electromechanical integration of cardiomyocytes derived from human embryonic stem cells. Nat Biotechnol 22:1282– 1289
21. Kikuchi K, McDonald AD, Sasano T, Donahue JK (2005) Targeted modification of atrial electrophysiology by homogeneous transmural atrial gene transfer. Circulation 111:264–270
22. Kirshenbaum LA, MacLellan WR, Mazur W, French BA, Schneider MD (1993) Highly efficient gene transfer into adult ventricular myocytes by recombinant adenovirus. J Clin Invest 92:381–387
23. Kolossov E, Lu Z, Drobinskaya I, Gassanov N, Duan Y, Sauer H, Manzke O, Bloch W, Bohlen H, Hescheler J, Fleischmann BK (2005) Identification and characterization of embryonic stem cell-derived pacemaker and atrial cardiomyocytes. FASEB J 19:577–579
24. Kyriakou CA, Yong KL, Benjamin R, Pizzey A, Dogan A, Singh N, Davidoff AM, Nathwani AC (2005) Human mesenchymal stem cells (hMSCs) expressing truncated soluble vascular endothelial growth factor receptor (tsFlk-1) following lentiviral-mediated gene transfer inhibit growth of Burkitt's lymphoma in a murine model. J Gene Med
25. Liechty KW, MacKenzie TC, Shaaban AF, Radu A, Moseley AM, Deans R, Marshak DR, Flake AW (2000) Human mesenchymal stem cells engraft and demonstrate site-specific differentiation after in utero transplantation in sheep. Nat Med 6:1282–1286
26. Miake J, Marbán E, Nuss HB (2002) Biological pacemaker created by gene transfer. Nature 419:132–133
27. Miake J, Marbán E, Nuss HB (2003) Functional role of in-ward rectifier current in heart probed by Kir2.1 overexpression and dominant-negative suppression. J Clin Invest 111:1529–1536

28. Michels G, Er F, Khan I, Sudkamp M, Herzig S, Hoppe UC (2005) Single-channel properties support a potential contribution of hyperpolarization-activated cyclic nucleotide-gated channels and If to cardiac arrhythmias. Circulation 111:399–404
29. Milanesi R, Baruscotti M, Gnecchi-Ruscone T, DiFrancesco D (2006) Familial sinus bradycardia associated with a mutation in the cardiac pacemaker channel. N Engl J Med 354:151–157
30. Mistrik P, Mader R, Michalakis S, Weidinger M, Pfeifer A, Biel M (2005) The murine HCN3 gene encodes a hyperpo-larization-activated cation channel with slow kinetics and unique response to cyclic nucleotides. J Biol Chem 280:27056–27061
31. Miyoshi H, Blomer U, Takahashi M, Gage FH, Verma IM (1998) Development of a self-inactivating lentivirus vector. J Virol 72:8150–8157
32. Muller OJ, Leuchs B, Pleger ST, Grimm D, Franz WM, Katus HA, Kleinschmidt JA (2006) Improved cardiac gene transfer by transcriptional and transductional targeting of adeno-associated viral vectors. Cardiovasc Res
33. Plotnikov AN, Sosunov EA, Qu J, Shlapakova IN, Anyukhovsky EP, Liu L, Janse MJ, Brink PR, Cohen IS, Robinson RB, Danilo P Jr, Rosen MR (2004) Biological pacemaker implanted in canine left bundle branch provides ventricular escape rhythms that have physiologically acceptable rates. Circulation 109:506–512
34. Potapova I, Plotnikov A, Lu Z, Danilo P Jr, Valiunas V, Qu J, Doronin S, Zuckerman J, Shlapakova IN, Gao J, Pan Z, Herron AJ, Robinson RB, Brink PR, Rosen MR, Cohen IS (2004) Human mesenchymal stem cells as a gene delivery system to create cardiac pacemakers. Circ Res 94:952–959
35. Qu J, Barbuti A, Protas L, Santoro B, Cohen IS, Robinson RB (2001) HCN2 overexpression in newborn and adult ventricular myocytes: distinct effects on gating and excitability. Circ Res 89:E8–E14
36. Qu J, Plotnikov AN, Danilo P Jr., Shlapakova I, Cohen IS, Robinson RB, Rosen MR (2003) Expression and function of a biological pacemaker in canine heart. Circulation 107:1106–1109
37. Robinson RB, Rosen MR, Brink PR, Cohen IS (2005) Letter regarding the article by Xue et al, ''Functional integration of electrically active cardiac derivatives from genetically engineered human embryonic stem cells with quiescent recipient ventricular cardiomyocytes''. Circulation 112:e82–e83
38. Rosen MR (2005) 15th annual Gordon K. Moe Lecture. Biological pacemaking: in our lifetime? Heart Rhythm 2:418–428
39. Rosen MR, Brink PR, Cohen IS, Robinson RB (2004) Genes, stem cells and biological pacemakers. Cardiovasc Res 64:12–23
40. Rosen MR, Robinson RB, Brink P, Cohen IS (2004) Recreating the biological pacemaker. Anat Rec A Discov Mol Cell Evol Biol 280:1046–1052
41. Seifert R, Scholten A, Gauss R, Mincheva A, Lichter P, Kaupp UB (1999) Molecular characterization of a slowly gating human hyperpolarization-activated channel predominantly expressed in thalamus, heart, and testis. Proc Natl Acad Sci USA 96:9391–9396
42. Shi W, Wymore R, Yu H, Wu J, Wymore RT, Pan Z, Robinson RB, Dixon JE, McKinnon D, Cohen IS (1999) Distribution and prevalence of hyperpolarization-activated cation channel (HCN) mRNA expression in cardiac tissues. Circ Res 85:e1–e6
43. Srivastava D, Ivey KN (2006) Potential of stem-cell-based therapies for heart disease. Nature 441:1097–1099
44. Stieber J, Thomer A, Much B, Schneider A, Biel M, Hofmann F (2003) Molecular basis for the different activation kinetics of the pacemaker channels HCN2 and HCN4. J Biol Chem 278:33672–33680

45. Su H, Joho S, Huang Y, Barcena A, Arakawa-Hoyt J, Grossman W, Kan YW (2004) Adeno-associated viral vector delivers cardiac-specific and hypoxia-inducible VEGF expression in ischemic mouse hearts. Proc Natl Acad Sci USA 101:16280–16285
46. Tan HL, Hou CJ, Lauer MR, Sung RJ (1995) Electrophysiologic mechanisms of the long QT interval syndromes and torsade de pointes. Ann Intern Med 122:701–714
47. Tang YL, Tang Y, Zhang YC, Qian K, Shen L, Phillips MI (2005) Improved graft mesenchymal stem cell survival in ischemic heart with a hypoxia-regulated heme oxygenase-1 vector. J Am Coll Cardiol 46:1339–1350
48. Thornton AS, Jordaens LJ (2006) Remote magnetic navigation for mapping and ablating right ventricular outflow tract tachycardia. Heart Rhythm 3:691–696
49. Toro R, Saadi I, Kuburas A, Nemer M, Russo AF (2004) Cell-specific activation of the atrial natriuretic factor promoter by PITX2 and MEF2A. J Biol Chem 279:52087–52094
50. Totsugawa T, Kobayashi N, Okitsu T, Noguchi H, Watanabe T, Matsumura T, Maruyama M, Fujiwara T, Sakaguchi M, Tanaka N (2002) Lentiviral transfer of the LacZ gene into human endothelial cells and human bone marrow mesenchymal stem cells. Cell Transplant 11:481–488
51. Tripathy SK, Black HB, Goldwasser E, Leiden JM (1996) Immune responses to transgene-encoded proteins limit the stability of gene expression after injection of replication-defective adenovirus vectors. Nat Med 2:545–550
52. Tristani-Firouzi M, Jensen JL, Donaldson MR, Sansone V, Meola G, Hahn A, Bendahhou S, Kwiecinski H, Fidzianska A, Plaster N, Fu YH, Ptacek LJ, Tawil R (2002) Functional and clinical characterization of KCNJ2 mutations associated with LQT7 (Andersen syndrome). J Clin Invest 110:381–388
53. Verma IM, Weitzman MD (2005) Gene therapy: twenty-first century medicine. Annu Rev Biochem 74:711–738
54. Xue T, Cho HC, Akar FG, Tsang SY, Jones SP, Marbán E, Tomaselli GF, Li RA (2005) Functional integration of electrically active cardiac derivatives from genetically engineered human embryonic stem cells with quiescent recipient ventricular cardiomyocytes: insights into the development of cell-based pacemakers. Circulation 111:11–20
55. Yellen BB, Forbes ZG, Halverson DS, Fridman G, Barbee KA, Chorny M, Levy R, Friedman G (2005) Targeted drug delivery to magnetic implants for therapeutic applications. J Magn Magn Mater 293:647–654
56. Zhang L, Benson DW, Tristani-Firouzi M, Ptacek LJ, Tawil R, Schwartz PJ, George AL, Horie M, Andelfinger G, Snow GL, Fu YH, Ackerman MJ, Vincent GM (2005) Electrocardiographic features in Andersen–Tawil syndrome patients with KCNJ2 mutations: characteristic T-U-wave patterns predict the KCNJ2 genotype. Circulation 111:2720–2726
57. Zhao J, Pettigrew GJ, Thomas J, Vandenberg JI, Delriviere L, Bolton EM, Carmichael A, Martin JL, Marber MS, Lever AM (2002) Lentiviral vectors for delivery of genes into neonatal and adult ventricular cardiac myocytes in vitro and in vivo. Basic Res Cardiol 97:348–358
58. Zufferey R, Dull T, Mandel RJ, Bukovsky A, Quiroz D, Naldini L, Trono D (1998) Self-inactivating lentivirus vector for safe and efficient in vivo gene delivery. J Virol 72: 9873–9880

Inhibition of Cardiomyocyte Automaticity by Electrotonic Application of Inward Rectifier Current from Kir2.1 Expressing Cells

Teun P. de Boer[1,4], Toon A.B. van Veen[1], Marien J.C. Houtman[1], John A. Jansen[1], Shirley C.M. van Amersfoorth[2,4], Pieter A. Doevendans[3], Marc A. Vos[1], and Marcel A.G. van der Heyden[1,4]

[1] Department of Medical Physiology, Heart Lung Center Utrecht, University Medical Center Utrecht, Yalelaan 50, 3584 Utrecht, The Netherlands
m.a.g.vanderheyden@umcutrecht.nl
[2] Experimental and Molecular Cardiology Group, Academic Medical Center, Amsterdam, The Netherlands
[3] Department of Cardiology, University Medical Center Utrecht, Utrecht, The Netherlands
[4] Interuniversity Cardiology Institute of the Netherlands, Utrecht, The Netherlands

Abstract. A biological pacemaker might be created by generation of a cellular construct consisting of cardiac cells that display spontaneous membrane depolarization, and that are electrotonically coupled to surrounding myocardial cells by means of gap junctions. Depending on the frequency of the spontaneously beating cells, frequency regulation might be required. We hypothesized that application of Kir2.1 expressing non-cardiac cells, which provide I_{K1} to spontaneously active neonatal cardiomyocytes (NCMs) by electrotonic coupling in such a cellular construct, would generate an opportunity for pacemaker frequency control. Non-cardiac Kir2.1 expressing cells were co-cultured with spontaneously active rat NCMs. Electrotonic coupling between the two cell types resulted in hyperpolarization of the cardiomyocyte membrane potential and silencing of spontaneous activity. Either blocking of gap-junctional communication by halothane or inhibition of I_{K1} by $BaCl_2$ restored the original membrane potential and spontaneous activity of the NCMs. Our results demonstrate the power of electrotonic coupling for the application of specific ion currents into an engineered cellular construct such as a biological pacemaker.

Keywords: Kir2.1, Inward rectifier, Cardiomyocyte, Pacemaker, Electrotonic coupling.

1 Introduction

Genetically engineered cell based pacemakers may become a valuable alternative for the current electronic pacemakers. The dominant endogenous pacemakers in the mammalian heart, such as the sinoatrial (SA) and atrioventricular nodes, constitute a specific spatially organized population of cardiac myocytes that have no primary contractile function. These cardiomyocytes are characterized by the

J.A.E. Spaan et al. (Eds.): Biopacemaking, BIOMED, pp. 94–104, 2007.
springerlink.com

expression of a specialized electronome, i.e., genes responsible for excitability and action potential propagation. This facilitates regular spontaneous depolarizations that by gradually increasing conductive properties toward the periphery of the node can drive working myocardium (for review see reference [1]). The four most relevant properties of the nodal cardiomyocyte electronome of most mammalians are the expression of pacemaker channels of the HCN type, the absence of fast inward sodium channels, the absence or minimal expression of inward rectifier channels of the Kir2.*x* type, and finally, the expression of specialized gap-junction protein isoforms (for review see reference [16]). Several studies have elaborated on the use of HCN ion channels in the construction of biological pacemakers, which is covered elsewhere in this issue. Here, we will further emphasize on the role of the inward rectifier potassium current for control of pacemaking frequency.

The inward rectifier current (I_{K1}) is predominantly active during the last phase of repolarization and subsequently phase 4 of the atrial and ventricular action potential, where it has its role in both establishing and stabilizing the resting membrane at a rather negative potential between –75 and –90 mV (for review see reference [2]). Inward rectification is established by increasing potassium conductivity at hyperpolarization, while the I_{K1} channels close upon membrane depolarizing. The molecular determinants of I_{K1} have convincingly been identified as the potassium ion channel constituents Kir2.1 (KCNJ2) and Kir2.2 (KCNJ12) [23], of which the former seems the predominant isoform in the heart. A tetramer of Kir2.*x* subunits will form one I_{K1} channel [14] either as a homomeric (Kir2.1 or Kir2.2 only) or heteromeric (combined Kir2.1 and Kir2.2) channel [24]. This molecular built-up enables an efficient dominant negative effect whenever one or more of the four subunits is modified and thereby disturbs the total I_{K1} channel conformation resulting in non-functional channels.

The absence of strong I_{K1} from the nodal tissues is a requirement for HCN ion channels to enable gradual depolarization by their pacemaker current (I_f), resulting in the nodal action potential. Consistent with the limited expression or even absence from the SA node, no increased sinus rhythm is seen in the Kir2.1 knock-out mouse [23]. Mice overexpressing a dominant negative form of Kir2.1 from the α-MHC promoter surprisingly displayed even a decrease in heart rate [8]. Neither ectopic beats nor reentry arrhythmias were observed in these animals, indicating redundancy for Kir2.1 loss in the intact murine myocardium. However, in isolated cardiomyocytes from both mice, an increased incidence of spontaneous action potentials was observed. In contrast to the two downregulation approaches of I_{K1}, its upregulation in transgenic mice overexpressing wildtype Kir2.1 in the heart resulted in the expected slowing in heart rate [6].

Mutations in Kir2.1 that result in less or non-functional I_{K1} channels may be considered as natural knock-outs. In recent years, several mutations have been identified, several of which behave as dominant negative, leading to a clinical manifestation known as Andersen syndrome [11, 20]. Amongst other features, Andersen syndrome patients may suffer from cardiac arrhythmias. Therefore, unlike mice, large animal species, including man, may be more vulnerable for a loss of I_{K1} from the heart with respect to the occurrence of arrhythmias. In principle, introduction of a dominant negative form of Kir2.1 in the working myocardium may undermine a stable resting membrane potential (RMP), and thereby increase vulnerability of the cell for membrane potential fluctuations, which may eventually result in a spontaneous depolarization. Dependent upon the size of such an unstable region, a real local pacemaker able to drive the heart may be formed. In guinea pig, where a more prominent role for Kir2.1 in ventricular cardiomyocyte membrane potential stabilization is suspected than in mice, a strong ectopic expression of dominant negative forms of Kir2.1 indeed leads to liberation of pacemaker activity in the ventricle [9].

Although introduction of potent dominant negative Kir2.1 may result in biological pacemaker activity, the disadvantage is its dependence on spontaneous membrane potential fluctuations rather than on pacemaker currents formed by the HCN family. A similar situation can be found in embryonic stem cell pacemaker action. During stem cell differentiation into cardiomyocytes, the developmental program is recapitulated resulting in spontaneously beating cells. In the P19 cell model, only a minority of the cells expresses functional I_f, and no I_{K1} is found in such cells [21] from which it can be concluded that pacemaker behavior occurs through RMP fluctuations allowed by the absence of the inward rectifier. Upon maturation, however, some cells start to express I_{K1}, which is accompanied by stable, more negative RMPs and a cessation of spontaneous beating. The uncontrolled maturation of embryonic stem cell derived cardiomyocytes makes them currently unsuitable for use as biological pacemakers. A promising avenue seems to use a cell-based tissue construct delivering HCN-mediated I_f to the surrounding myocardium [12]. Such a biological pacemaker cell construct needs to be tightly controlled, and we hypothesized that biologically engineered ion channel expressing cells could attribute to biological pacemaker regulation through electrotonic coupling. Here, we demonstrate, as a proof of principle, that I_{K1} expressing HEK-293 cells can modulate the beating frequency of spontaneously active neonatal rat cardiomyocytes through electronic interaction.

2 Methods

2.1 HEK-KWGF Cells and Neonatal Cardiomyocytes

HEK-293 cells (ATCC # CRL-1573) were regularly cultured in DMEM supplemented with 10% fetal calf serum, 2 mM L-glutamine, 50 U/ml penicillin, and 50 μg/ml streptomycin, all purchased from Cambrex (Verviers, Belgium).

HEK-293 cells stably expressing murine wildtype Kir2.1-GFP (HEK-KWGF) were generated as follows. HEK-293 cells were transfected with a pcDNA3-Kir2.1- GFP, producing a C-terminal GFP-tagged Kir2.1 (a generous gift of Anatoli Lopatin [8]) using Lipofectamin (InVitrogen, Paisly, UK) and selected for stable transfectants with 500 μg/ml G418 (Sigma, St. Louis, MI, USA) for 3 weeks. Complete stable pools of Kir2.1-GFP expressing cells were FACS sorted. The upper 10% of the total population displaying the strongest GFP expression was isolated, proceeded in culture, and denominated as HEK-KWGF cells.

Neonatal rat cardiomyocytes (NCMs) were isolated and cultured as described before [19] in accordance with the Guide for the Care and Use of Laboratory Animals published by the US National Institutes of Health (NIH Publication No. 85–23, revised 1996) and was approved by the institutional committee for animal experiments, with only slight modifications. Sixty micrograms per milliliter Pancreatine (Sigma) was used instead of 20 μg/ml DNase, cells were resuspended in culture medium consisting of M199 with Hanks, HEPES, L-AA and L-glutamine (Gibco, Breda, The Netherlands) and 10% neonatal bovine serum (Gibco), and preplating was shortened to 2 h. Co-cultures were made by plating NCM on the day of isolation and on the next day, HEK-KWGF cell suspensions were added at the desired concentrations.

2.2 Western Blot Analysis

Cells were lysed in RIPA buffer (20 mM Tris–HCl, pH 7.4, 150 mM NaCl, 10 mM Na_2HPO_4, 1% (v/v) Triton X-100, 1% (w/v) Na-deoxycholate, 0.1% (w/v) SDS, 1 mM EDTA, 50 mM NaF, 1 mM PMSF, 10 μg/ml aprotinin). Lysates were clarified by centrifugation at 14,000g for 5 min at 4°C. Twenty micrograms of protein lysate was mixed with Laemmli sample buffer and proteins were separated by 10% SDS-PAGE and subsequently electro-blotted onto nitrocellulose membrane (Biorad, Veenendaal, The Netherlands). Antibodies used: connexin43 (Cx43) (Zymed, San Francisco, CA, USA; cat. nr. 71-0700), Pan-Cadherin (Sigma, St Louis, MI, USA; cat. nr. C-3678), and GFP (Santa Cruz Biotechnology, Santa Cruz, CA, USA; cat. nr. SC-9996). Proteins were visualized by using peroxidase labeled secondary antibody (Jackson, Soham, UK) and standard ECL procedures (Amersham Bioscience, Roosendaal, The Netherlands).

2.3 I_{K1} and Action Potential Recording

A HEKA EPC-7 patch clamp amplifier was used to measure electrotonic interaction between cells. Voltage and current signals were recorded using a custom data acquisition program (kindly provided by Dr J.G. Zegers) running on an Apple Macintosh computer equipped with a 12-bit National Instruments PCI-MIO-16E-4 acquisition card. Current signals were low-pass filtered at 2.5 kHz and acquired at 10 kHz. Action potentials were elicited with a brief square current pulse. I_{K1} was recorded in voltage clamp mode of the whole-cell patch clamp configuration. A conventional mono-exponential fitting procedure was used to derive membrane capacitance from currents elicited by - 10 and +10 mV square pulses while holding the cell at RMP. Series resistance was not compensated for. I_{K1} was elicited by applying 1 s square test pulses ranging between - 130 and + 50 mV from a holding potential of –40 mV. Steady state currents at the end of the pulse were normalized to membrane capacitance and plotted against test pulse potential. Offline analysis was done using MacDaq 8.0 (kindly provided by Dr A.C.G. van Ginneken) and R 2.0.1 [13].

Experiments were done at 20° C on a Nikon Diaphot 300 inverted microscope. Extracellular buffer used was a modified Tyrode's solution, containing (in mM) NaCl 140, KCl 5.4, $CaCl_2$ 1.8, $MgCl_2$ 1, HEPES 15, $NaHCO_3$ 35, glucose 6, pH 7.20/NaOH. Gap-junctional coupling was inhibited by applying 4 mM Halothane dissolved in extracellular buffer at the interface between NCM and HEK-KWGF cells with an additional micropipette. Pipette buffer contained (in mM) potassium gluconate 125, KCl 10, HEPES 5, EGTA 5, $MgCl_2$ 2, $CaCl_2$ 0.6, Na_2ATP 4, pH 7.20/KOH. Patch pipettes were pulled on a Narishige PC-10 puller and fire-polished. When filled with pipette buffer, the pipette resistance ranged between 2 and 5 MΩ. Liquid junction potential was calculated using Clampex (Molecular Devices Corp, Sunnyvale, CA, USA) and used for offline correction.

Beating frequency was determined by counting the number of beats for a period of 30 s under visual inspection. For each condition, at least six independent NCM or NCM/HEK-KWGF clusters were analyzed.

3 Results

3.1 Validation of I_{K1} Expressing HEK-KWGF Cells

To create a co-culture system able to test electrotonic coupling of I_{K1} expressing cells with spontaneous active NCMs, we first created cells which were in principle able to couple with cardiomyocytes and that expressed functional I_{K1}. Therefore, HEK-293 cells were stably transfected with a murine Kir2.1-GFP fusion protein expression construct. Upon G418 based selection and FACS assisted enrichment, the resulting HEK-KWGF cells were further characterized. Western blot analysis demonstrated the expression of the Kir2.1-GFP fusion protein in stably transfected cells, while no signal was observed in maternal HEK-293 cells or NCM (Fig. 1a). Analysis of the gap-junction protein Cx43 expression demonstrated the presence of Cx43 in HEK-293 and HEK-KWGF cells; however, the levels were fairly low when compared to Cx43 expression in rat NCMs (Fig. 1a). In contrast, Cadherin expression, required for mechanical interaction between the two cell types, displayed less difference in expression level between the cell types. Epifluorescence microscopy illustrated the presence of a strong GFP signal at the cell borders of HEK-KWGF cells, and therefore Kir2.1 channels appear to be expressed at the plasma membrane (Fig. 1b). Subsequently, electrophysiological measurements indeed validated the presence

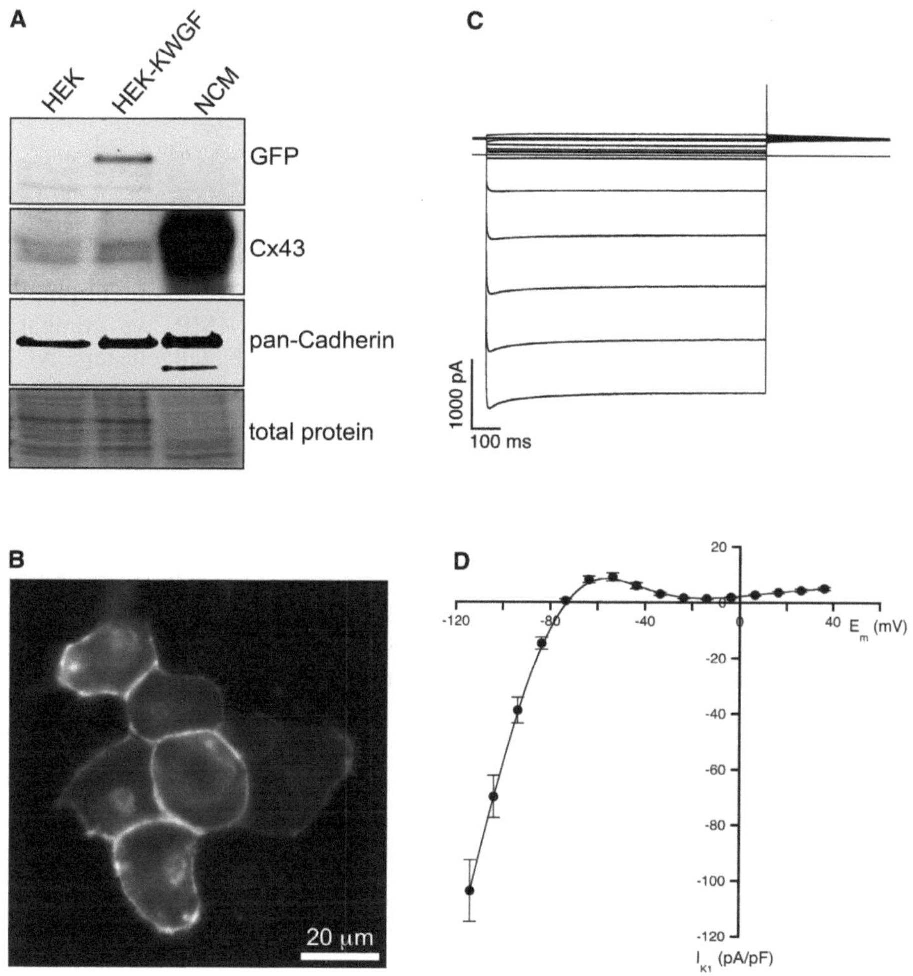

Fig. 1. Characterization of HEK-KWGF cells. **a** Western blot analysis of non-transfected HEK-293 cells (HEK), Kir2.1-GFP expressing HEK-293 cells (HEK-KWGF) and rat neonatal cardiomyocytes (NCMs). Kir2.1-GFP signal (GFP) was only observed in HEK-KWGF cells. Connexin43 is expressed at low levels in HEK and HEK-KWGF cells, and at high levels in NCM. Cadherin adherence junction proteins (pan-Cadherin) are expressed in all three cell types at similar levels. Total protein staining is used as loading control. **b** GFP fluorescence microscopy displays strong plasma membrane localization of Kir2.1-GFP. **c** Current traces of HEK-KWGF cells characteristic for I_{K1}. Scale bars—horizontal = 100 ms, vertical = 1,000 pA. **d** Current–voltage relationship of HEK-KWGF cells displaying I_{K1} characteristics

of high levels of functional I_{K1} channels in these cells, displaying an I–V curve characteristic for Kir2.1 channels (Fig. 1c, d). Current densities were −58.7 ± 6.6 pA/pF at −100 mV, which is approximately 6–15 fold higher than in isolated neonatal (approximately −4 pA/pF) [23] and adult murine heart cells

(–4.68 and –9.3 pA/pF) [6, 8]. In rat NCMs, I_{K1} current densities between - 7 and – 9 pA/pF at –100 mV have been reported [5, 7]. HEK-KWGF cells had a mean membrane potential of - 75.2 ± 0.5 mV (N = 20). No I_{K1} was found in non-transfected HEK- 293 cells.

3.2 Co-culture of HEK-KWGF Cells with NCMs

Neonatal rat cardiomyocytes share several features with nodal cardiomyocytes, of which their automaticity is the most characteristic. Furthermore, they display a maximal diastolic membrane potential of –65 mV and a prominent phase 4 depolarization. As I_{K1} is the main determinant of the RMP, we compared RMP of NCM clusters with that of NCM/HEK-KWGF cell clusters. NCM clusters displayed a maximal diastolic membrane potential of –65.1 ± 2.8 mV (N = 4), which significantly decreased to –77.6 ± 3.4 mV (N = 6) ($p < 0.01$; Student's *t*-test) in NCMs coupled to one or more HEK-KWGF cells. When attached to HEK-KWGF, no spontaneous beating of NCMs was observed, nor could action potentials be recorded from those NCMs. Upon triggering by a depolarizing current, however, action potentials could be elicited that displayed an aberrant morphology when compared to monocultures of NCMs (Fig. 2a). Application of the gap-junction blocker Halothane to NCM/HEK-KWGF cell pairs decreased RMP of the NCM to its normal value and restored spontaneous action potential formation and beating behavior (Fig. 2b).

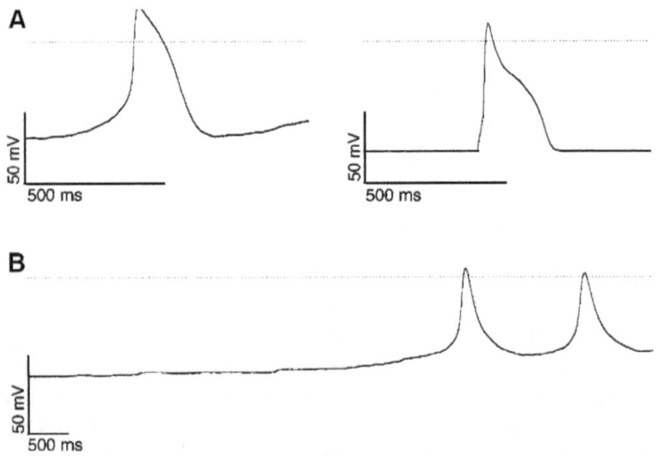

Fig. 2. Electrotonic coupling between HEK-KWGF and neonatal cardiomyocyte (NCM). **a** Action potential recordings from spontaneous NCM mono-cultures (*left panel*) and triggered NCM/HEK-KWGF cell pair (*right panel*) as measured from the NCM. Note the diffrrence in resting membrane potential (RMP), action potential upstroke and duration. **b** Gap-junction uncoupling by halothane restores RMP and spontaneous action potentials in NCM/HEK-KWGF cell pair as measured from the NCM. Scale bars—horizontal = 500 ms, vertical = 50 mV, *dashed line* indicates 0 mV.

We next questioned whether HEK-KWGF cells could be used in co-cultures to control beating frequency, i.e., pacing rate, of a NCM culture. Therefore, co-cultures containing 300,000 cells were made with 0, 5, 25, 50 or 75% contribution of HEK-KWGF cells. NCM monocultures displayed a beating rate of approximately 238 bpm (Fig. 3). Co-culture with 5% HEK-KWGF cells dramatically decreased the beating rate to 83 bpm. Increasing the amount of HEK-KWGF cells inhibited beating rates even further. As it has been shown that I_{K1} can be inhibited by 1 mM $BaCl_2$ [17] (Fig 3, inset), we tested whether addition of $BaCl_2$ to the co-cultures could rescue the original beating rates. In NCM mono-cultures, $BaCl_2$ application resulted in a slight increase (12%) in beating rate, suggesting the presence of limited amounts of endogenous I_{K1}. When applied to co-cultures, $BaCl_2$ could strongly inhibit the silencing effect of HEK-KWGF with respect to NCM beating rate (Fig. 3).

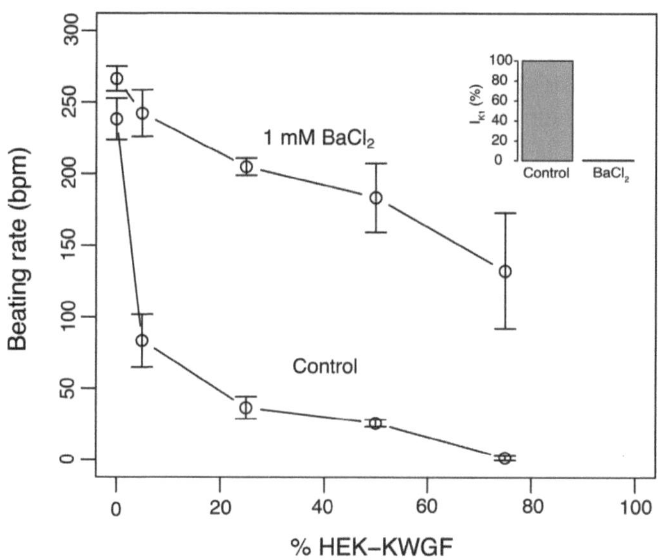

Fig. 3. I_{K1} dependent regulation of neonatal cardiomyocyte (NCM) beating frequency. NCM and HEK-KWGF cells were co-cultured at a density of 300,000 cells/well with an increasing percentage of HEK-KWGF with respect to NCM cells. Beating frequencies (bpm) of NCM in the co-cultures were determined in the absence (control) and presence of 1 mM BaCl2 (BaCl2). Results from six independent measurements ± SEM are given. Significant differences (control vs BaCl2) ($p < 0.01$) were found at all points except for 0% HEK-KWGF, and beating rate at 25% HEK-KWGF cells were statistically different ($p < 0.01$) from 0 to 5% HEK-KWGF cells within each condition (two-way ANOVA). Inset: 99% inhibition of peak I_{K1} by BaCl2 was determined in HEK-KWGF cells (N = 4).

4 Discussion

Our study demonstrates a hyperpolarizing effect of Kir2.1 on the RMP of neonatal cardiac myocytes by electrotonic coupling, which thereby results in inhibition of spontaneous depolarizations of the cardiomyocytes. A remarkably strong inhibition of the beating frequency was accomplished with relatively few HEK-KWGF cells (Fig. 3). This may be explained by the large difference in I_{K1} densities between the two cell types, which are –60 and –8 pA/pF at –100 mV for HEK-KWGF and NCM cells, respectively. This mechanism would be of support to create cellular constructs consisting of a heterogeneous population of I_f expressing cells and I_{K1} expressing cells, which dependent on their relative contribution within the construct might generate a predetermined pacemaker frequency. This approach would obviously benefit from a controlled expression of I_{K1}. This could be reached by generating clonal cells expressing different amounts of Kir2.1 and thereby I_{K1} Alternatively, intervention in natural Kir2.1 regulatory mechanisms could be used, such as PKC or PKA mediated effects on I_{K1} [4, 15]. Furthermore, Kir2.1 channels appear sensitive to intracellular polyamine [10] or PIP_2 regulation [18] which seems to operate at least in part interdependently [22]. Obvious drawback of these approaches is the rather limited specificity of the signaling interference which may have many undesired effects on other ion channels in the heart or on I_{K1} in other tissues than the heart. More specific, and thus in favor, is to produce cells in which Kir2.1 is placed under the control of an inducible promoter, in which the inducing agent has no other physiologically relevant biological function. To achieve this, ecdysone could be an inducer of choice [3].

The current promises in cardiac tissue engineering aside, the preclinical research now first has to establish the applicability of proofs of principles studies like this one, for their eventual clinical relevance.

5 Conclusion

Electrotonic coupling of Kir2.1 expressing cells to spontaneous depolarizing NCMs results in silencing of the spontaneous beating behavior of the latter. This system is a proof of concept demonstrating the power of electrotonic coupling for the application of specific ion currents into an engineered cellular construct such as a biological pacemaker.

Acknowledgments. We thank Anatoli Lopatin for sharing Kir2.1-GFP expression construct and Henk Rozemuller for FACS sorting of the HEK-KWGF cells. This study is supported by the Technology Foundation (STW program DPTE, grant #MKG5942, MvdH and grant UGT.6746, TvV), the Netherlands Heart Foundation (grant 2003B073, TdB) and the Netherlands Organization for Scientific Research (NWO, grant 916.36.012, TvV). FP6 (Framework Program LSHB-CT-2004-502988) of the European Committee (BK).

References

1. Boyett MR, Honjob H, Kodama I (2000) The sinoatrial node, a heterogeneous pacemaker structure. Cardiovasc Res 47:658–687. DOI 10.1016/S0008-6363(00)00135-8
2. Dhamoon AS, Jalife J (2005) The inward rectifier current (I_{K1}) controls cardiac excitability and is involved in arrhythmogenesis. Heart Rhythm 2:316–324. DOI 10.1016/ j.hrthm.2004.11.012
3. Harvey DM, Caskey CT (1998) Inducible control of gene expression: prospects for gene therapy. Curr Opin Chem Biol 2:512–518. DOI 10.1016/S1367-5931(98) 80128-2
4. Koumi S, Backer CL, Arentzen CE, Sato R (1995) β-Adrenergic modulation of the inwardly rectifying potassium channel in isolated human ventricular myocytes. Alteration in channel response to β-adrenergic stimulation in failing human hearts. J Clin Invest 96:2870–2881
5. Lange PS, Er F, Gassanov N, Hoppe UC (2003) Andersen mutations of KCNJ2 suppress the native inward rectifier current I_{K1} in a dominant-negative fashion. Cardiovasc Res 59:321–327. DOI 10.1016/S0008-6363(03)00434-6
6. Li J, McLerie M, Lopatin A (2004) Transgenic upregulation of I_{K1} in the mouse heart leads to multiple abnormalities of cardiac excitability. Am J Physiol Heart Circ Physiol 287:H2790– H2802. DOI 10.1152/ajpheart.00114.2004
7. Masuda H, Sperelakis N (1993) Inwardly rectifying potassium currents in rat fetal and neonatal ventricular cardiomyocytes. Am J Physiol Heart Circ Physiol 265:H1107–H1111
8. McLerie M, Lopatin A (2003) Dominant-negative suppression of I_{K1} in the mouse heart leads to altered cardiac excitability. J Mol Cell Cardiol 35:367–378. DOI 10.1016/S0022- 2828(03)00014-2
9. Miake J, Marbán E, Nuss HB (2002) Gene therapy: biological pacemaker created by gene transfer. Nature 419:132–133. DOI 10.1038/419132b
10. Panama BK, Lopatin AN (2006) Differential polyamine sensitivity in inwardly rectifying Kir2 potassium channels. J Physiol 571:287–302. DOI 10.1113/jphysiol.2005.097741
11. Plaster NM, Tawil R, Tristani-Firouzi M, Canú n S, Bendahhou S, Tsunoda A, Donaldson MR, Iannaccone ST, Brunt E, Barohn R, Clark J, Deymeer F, George AL Jr, Fish FA, Hahn A, Nitu A, Ozdemir C, Serdaroglu P, Subramony SH, Wolfe G, Fu Y-H, Ptáček LJ (2001) Mutations in Kir2.1 cause the developmental and episodic electrical phenotype of Andersen's syndrome. Cell 105:511–519. DOI 10.1016/S0092-8674(01)00342-7
12. Potapova I, Plotnikov A, Lu Z, Danilo P Jr, Valiunas V, Qu J, Doronin S, Zuckerman J, Shlapakova IN, Gao J, Pan Z, Herron AJ, Robinson RB, Brink PR, Rosen MR, Cohen IS (2004) Human mesenchymal stem cells as a gene delivery system to create cardiac pacemakers. Circ Res 94:952–959. DOI 10.1161/01.RES.0000123827.60210.72
13. R Development Core Team (2005) R: a language and environment for statistical computing. R Foundation for Statistical Computing, Vienna, Austria. URL http://www.R- project.org
14. Raab-Graham KF, Vandenberg CA (1998) Tetrameric subunit structure of the native brain inwardly rectifying potassium channel Kir2.2. J Biol Chem 273:19699–19707. DOI 10.1074/ jbc.273.31.19699

15. Sato R, Koumi S (1995) Modulation of the inwardly rectifying K^+ channel in isolated human atrial myocytes by α-1-adrenergic stimulation. J Membr Biol 148:185–191. DOI 10.1007/ BF00207274
16. Satoh H (2003) Sino-atrial nodal cells of mammalian hearts: ionic currents and gene expression of pacemaker ionic channels. J Smooth Muscle Res 39:175–193. DOI 10.1540/jsmr.39.175
17. Schram G, Melnyk P, Pourrier M, Wang Z, Nattel S (2002) Kir2.4 and Kir2.1 K^+ channel subunits co-assemble: a potential new contributor to inward rectifier current heterogeneity. J Physiol 544:337–349. DOI 10.1113/jphysiol.2002.026047
18. Takano M, Kuratomi S (2003) Regulation of cardiac inwardly rectifying potassium channels by membrane lipid metabolism. Prog Biophys Mol Biol 81:67–79. DOI 10.1016/S0079- 6107(02)00048-2
19. Teunissen BEJ, Van Amersfoorth SCM, Opthof T, Jongsma HJ, Bierhuizen MFA (2002) Sp1 and Sp3 activate the rat connexin40 proximal promoter. Biochem Biophys Res Commun 292:71–78. DOI 10.1006/bbrc.2002.6621
20. Tristani-Firouzi M, Jensen JL, Donaldson MR, Sansone V, Meola G, Hahn A, Bendahhou S, Kwiecinski H, Fizianska A, Plaster N, Fu YH, Ptáček LJ, Tawil R (2002) Functional and clinical characterization of KCNJ2 mutations associated with LQT7 (Andersen syndrome). J Clin Invest 110:381–388. DOI 10.1172/ JCI200215183
21. Van der Heyden MAG, Van Kempen MJA, Tsuji Y, Rook MB, Jongsma HJ, Opthof T (2003) P19 embryonal carcinoma cells: a suitable model system for cardiac electrophysiological differentiation at the molecular and functional level. Cardiovasc Res 58:410–422. DOI 10.1016/S0008-6363(03)00247-5
22. Xie LH, John SA, Ribalet B, Weiss JN (2005) Long polyamines act as cofactors in PIP_2 activation of inward rectifier potassium (Kir2.1) channels. J Gen Physiol 126:541–549. DOI 10.1085/ jgp.200509380
23. Zaritsky JJ, Redell JB, Tempel BL, Schwarz TL (2001) The consequences of disrupting cardiac inwardly rectifying K^+ current (I_{K1}) as revealed by the targeted deletion of the murine Kir2.1 and Kir2.2 genes. J Physiol 533:697–710. DOI 10.1111/j.1469-7793.2001.t01-1-00697.x
24. Zobel C, Cho HC, Nguyen T-T, Pekhletski R, Diaz RJ, Wilson GJ, Backx PH (2003) Molecular dissection of the inward rectifier potassium current (I_{K1}) in rabbit cardiomyocytes: evidence for heteromeric co-assembly of Kir2.1 and Kir2.2. J Physiol 550:365–372. DOI 10.1113/jphysiol.2002.036400

Propagation of Pacemaker Activity

Ronald W. Joyner[1], Ronald Wilders[2], and Mary B. Wagner[1]

[1] Department of Pediatrics, Emory University
School of Medicine, 2015 Uppergate Drive NE,
Atlanta, GA 30322, USA
rjoyner@cellbio.emory.edu
[2] Department of Physiology, Academic Medical Center,
Amsterdam, The Netherlands

Abstract. Spontaneous activity of specific regions (e.g., the Sinoatrial node, SAN) is essential for the normal activation sequence of the heart and also serve as a primary means of modulating cardiac rate by sympathetic tone and circulating catecholamines. The mechanisms of how a small SAN region can electrically drive a much larger atrium, or how a small ectopic focus can drive surrounding ventricular or atrial tissue are complex, and involve the membrane properties and electrical coupling within the SAN or focus region as well as the membrane properties, coupling conductance magnitudes and also regional distribution within the surrounding tissue. We review here studies over the past few decades in which mathematical models and experimental studies have been used to determine some of the design principles of successful propagation from a pacemaking focus. These principles can be briefly summarized as (1) central relative uncoupling to protect the spontaneously firing cells from too much electrotonic inhibition, (2) a transitional region in which the cell type and electrical coupling change from the central SAN region to the peripheral atrial region, and (3) a distributed anisotropy to facilitate focal activity.

Keywords: Sinoatrial node Electrical coupling Action potential propagation Cardiac rhythm Cardiac cell electrophysiology Numerical simulations.

1 Introduction

There are several regions of the heart in which spontaneous generation of action potentials can potentially occur. These include the normally dominant pacemaker region of the sinoatrial node but also regions such as the atrioventricular node and the Purkinje system whose intrinsic automaticity is normally suppressed by the faster automaticity of the sinoatrial node area. In addition, other regions of the heart may, under pathological conditions, also demonstrate automaticity which may lead to single ectopic beats, initiation of reentrant tachycardia, or persistent focal tachycardia [1,11]. One common feature of both normal and abnormal manifestations of automaticity is the propagation of the action potential from the automatic region to the quiescent surrounding atrial or ventricular tissue. While the presence of a coupling conductance between the automatic focus and the surrounding cells is necessary for propagation out from the focus region, this coupling conductance may also suppress the activity of the focus region by electrotonic interactions during the diastolic depolarization phase of the focus cells. This loading effect depends on many factors,

J.A.E. Spaan et al. (Eds.): Biopacemaking, BIOMED, pp. 105–120, 2007.
springerlink.com

including the size (number of cells) of the focus region, the intercellular coupling among the focus cells, the input resistance of the surrounding quiescent cells, and the value and spatial orientation of the coupling conductances among the surrounding cells. Another factor that is less well appreciated is that the focus region itself may be inhomogeneous both in terms of membrane properties as well as in the distribution of cellular coupling [7,8]. In this review, we will illustrate some of the experimental and theoretical studies, which have been done to determine some of the interactions among these factors.

2 Interactions Among Spontaneously Pacing Cells

When cells with intrinsically different spontaneously pacing rates are electrically connected, they might be expected to in some way ''synchronize'' their pacing to form a better current source for activation of surrounding quiescent tissue. A number of simulation and experimental studies have been done on this phenomenon. We did some early simulations [13] with the Noma–Irisawa [10] model of spontaneously pacing SA nodal cells in which we assumed two populations of cells: one with the normal model properties and another group with increased *L*-type calcium current ($2 \times I_{si}$) such that their automaticity was increased. In this case we considered these to be two aggregates of cells (similar to the chick embryo cell aggregates experimentally studied later on by Veenstra and DeHaan [19]) with each model aggregate having a membrane area of 1 mm^2. As shown in Fig. 1 the resulting cycle length (CL) for low coupling resistance is the same as if there were a single aggregate with an average value of 1.5 times the normal I_{si}. However, as the coupling resistance between the aggregates was increased, the synchronized CL increased toward the value of the faster aggregate. As the synchronized CL increased there was also a greater delay between the activation times of the two aggregates and a greater disparity in the maximum d*V*/d*t* of the upstrokes of the action potentials. The effects of increased coupling resistance (e.g., fewer gap junctions) can be thought of as converting the process of consensual synchronization to that of local propagation in which the cells with greater intrinsic automaticity thus dominate the overall rate. Note also that the ability of the two aggregates to synchronize was preserved up to a coupling resistance of 25 MΩ, even though the surface area of the model aggregate (1×10^{-2} cm^2) was much larger than a single cell. From the approximate membrane surface area of only 20×10^{-6} cm^2 for single nodal cells, we could extrapolate from these results that a resistance of up to 12.5×10^9 Ω would allow synchronization of a pair of nodal cells. This corresponds to an intercellular coupling conductance of ~0.1 nS which could be produced by a very few gap junction channels.

More recently, we used our ''coupling clamp'' circuit [24] to actually couple together isolated rabbit SA nodal cells with a defined coupling conductance [20]. In this technique, we simultaneously record the potential of two cells physically isolated from each other and use a computer system to calculate what coupling current would exist if there were a specified coupling conductance (G_c) and then, updated at short time intervals (e.g., 25 μs), supply this time varying current to one cell and the negative of this current to the other cell. Our results are shown in the top panel of Fig. 2a.

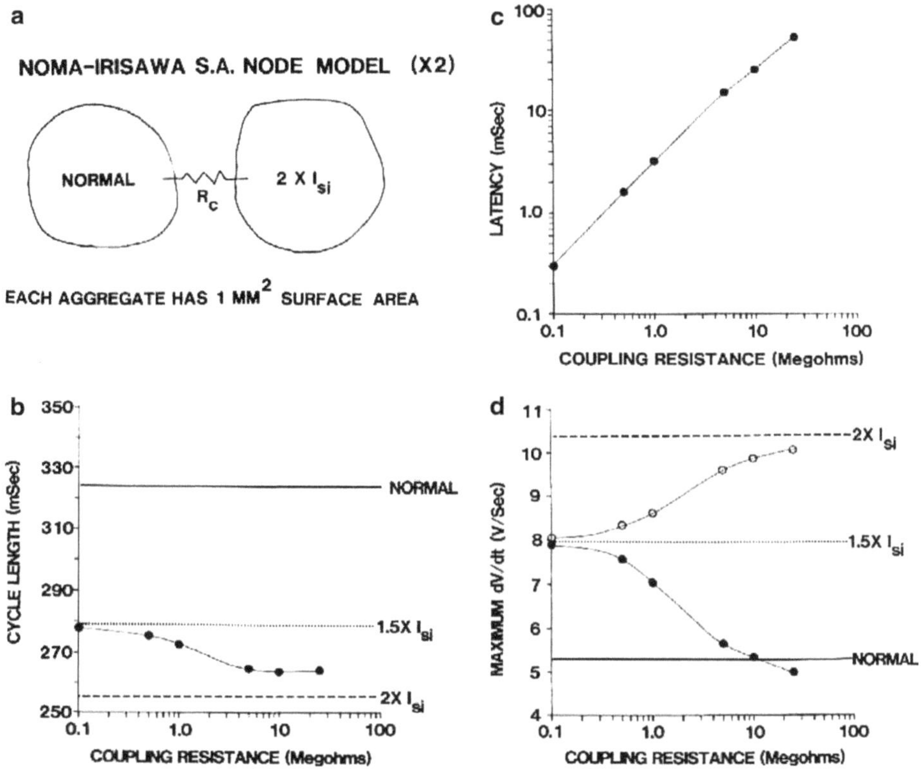

Fig. 1. Interactions between two coupled regions with intrinsically different pacing frequencies. **a** is a diagram of the simulation system for two isolated (groups of) cells with a coupling resistance R_c. **b** shows the relationship between coupling resistance and the entrained CL. The horizontal lines indicate the intrinsic CLs for isolated cells with the normal model (*solid line*) and the model with two times the normal L type Calcium current (I_{si})(*dashed line*) and the model with 1.5 times the normal I_{si} (*dotted line*). **c** shows the latency between action potentials occurring in the two groups of cells as a function of the coupling resistance. **d** shows the individual maximum d*V*/d*t* (V_{max}) values for the two groups of cells as functions of coupling resistance. The *open circles* are for the aggregate with the two times normal I_{si} model. The *filled circles* are for the aggregate with the normal model. As in (**b**), the *horizontal lines* indicate the intrinsic V_{max} for the models as labeled. From Joyner et al. [13].

During the periods of uncoupling the spontaneous activity of cell A is occurring at a shorter interbeat interval (IBI) (310 ms) than the spontaneous activity of cell B (390 ms). The action potentials of the two cells are also somewhat different in shape, with cell A (*solid line*) having a less negative maximum diastolic potential (–57 mV vs. –62 mV) and a less positive peak amplitude (26 mV vs. 28 mV) than cell B (*dotted line*). Cell A also has a shorter action potential duration than cell B. The lower panel of Fig. 2a plots the coupling current for this cell pair. The coupling current is, of course, zero during the two periods of uncoupling, and is plotted as a positive current in the direction from cell A to cell B. When the cells are coupled even with this small coupling conductance (G_c = 0.2 nS) a stable pattern of entrainment of the action potentials of cell A and the action potentials of cell B is established during the period of coupling. In Fig. 2 the action potentials of cell A and cell B are entrained at a common IBI, but the shapes of the action potentials are still quite different for cell A and cell B, with cell B retaining a more negative maximum diastolic potential and a longer action potential duration. In other words, the cells show "frequency entrainment," but not "waveform entrainment" [4]. When the coupling conductance was further increased to 10 nS (not shown) the action potentials of cell A and cell B are nearly synchronous during the period of coupling with the action potentials shapes now also nearly identical: the cells show both frequency and waveform entrainment.

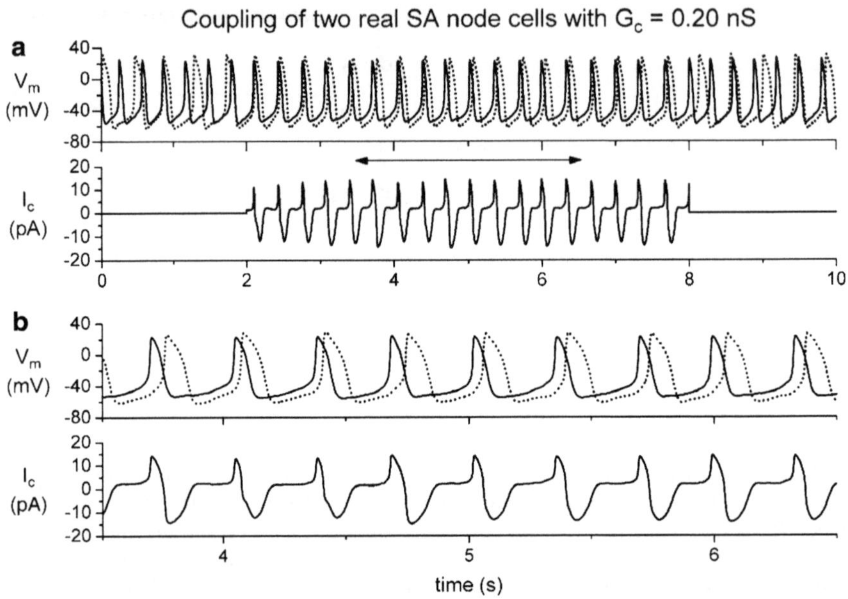

Fig. 2. Simultaneous recording for 10 s of two isolated rabbit sinoatrial node cells, with the cells uncoupled during the first 2 s and the last 2 s and coupled with a coupling conductance of 0.2 nS during the central 6 s. **a** Membrane potential (V_m; *upper panel*) of cell A (*solid line*) and cell B (*dotted line*), and coupling current (I_c; lower panel). **b** Data of part a replotted for the time period indicated by the *horizontal two-headed arrow* of **a**. From Verheijck et al. [20].

If we define the successive activation times of either cell A or cell B as the times at which the membrane potential crosses zero in a positive direction, we can analyze the effects of the coupling conductance on the time-varying IBIs and the activation delays between cell A and cell B for coupled action potentials. Figure 3 illustrates the effects of coupling conductance of 0.2 nS for the same cell pair used for Fig. 2. The two cells show fluctuations of their IBI during the uncoupled periods, as previously demonstrated for isolated spontaneously active cells [22] with a coefficient of variation for IBI of about 4%, but during the coupling period the IBI of the two cells has a mean value slightly larger than that of the uncoupled IBI of cell A. Note that the common IBI is clearly not the arithmetic average of the IBI values for each of the cells. Fluctuations of IBI still occur during the coupling period with differences of up to 34 ms for IBI values of the two cells even for cycles, which are associated in time as part of the entrainment process. The activation delay (lower panel) shows significant fluctuations from 24 to 64 ms with an average value of 43 ms.

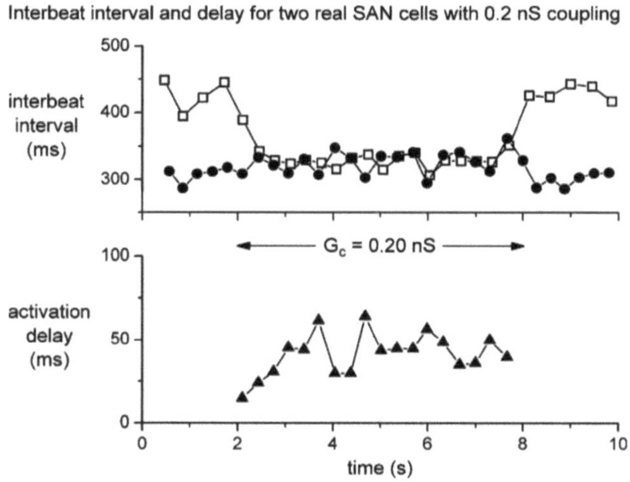

Fig. 3. Interbeat interval for the two rabbit SAN cells of Fig. 2 (*upper panel*; *filled circles* and *open squares*, respectively) and the delay in activation between the two cells (*lower panel*; *filled triangles*) for a coupling conductance G_c = 0.2 nS (indicated by a *horizontal arrow*). From Verheijck et al. [20].

3 Interactions Among a Spontaneously Pacing Cell Model and an Atrial Cell Model

For the simulations shown in Fig. 1, we used two isopotential spontaneously active membrane regions coupled by a resistance R_c. For the simulations of Fig. 4 we used an SA NODE region with the same Irisawa–Noma [10] model as for Fig. 1, and for the ATRIAL region we used the model of Beeler and Reuter [2] with the only modification being a reduction in the magnitude of the conductance for the slow inward current to 20% of the standard model to produce an action potential duration of ~100 ms to correspond with the atrial action potential duration [14]. It would have been more appropriate to also decrease the I_{K1} of the Beeler–Reuter model to match the input resistance of atrial cells (see section E) so these simulations somewhat overestimate the effects of electrical load of the surrounding cells on the function of the SAN region. These results [14] illustrate the three possibilities of such a coupled system. Each symbol plotted represents a simulation of 2 s, with the simulation results being either (a) failure of the SA region to pace (NOT PACE, *solid circle*), (b) pacing of the SA region without activation of the atrial region (PACE BUT NOT

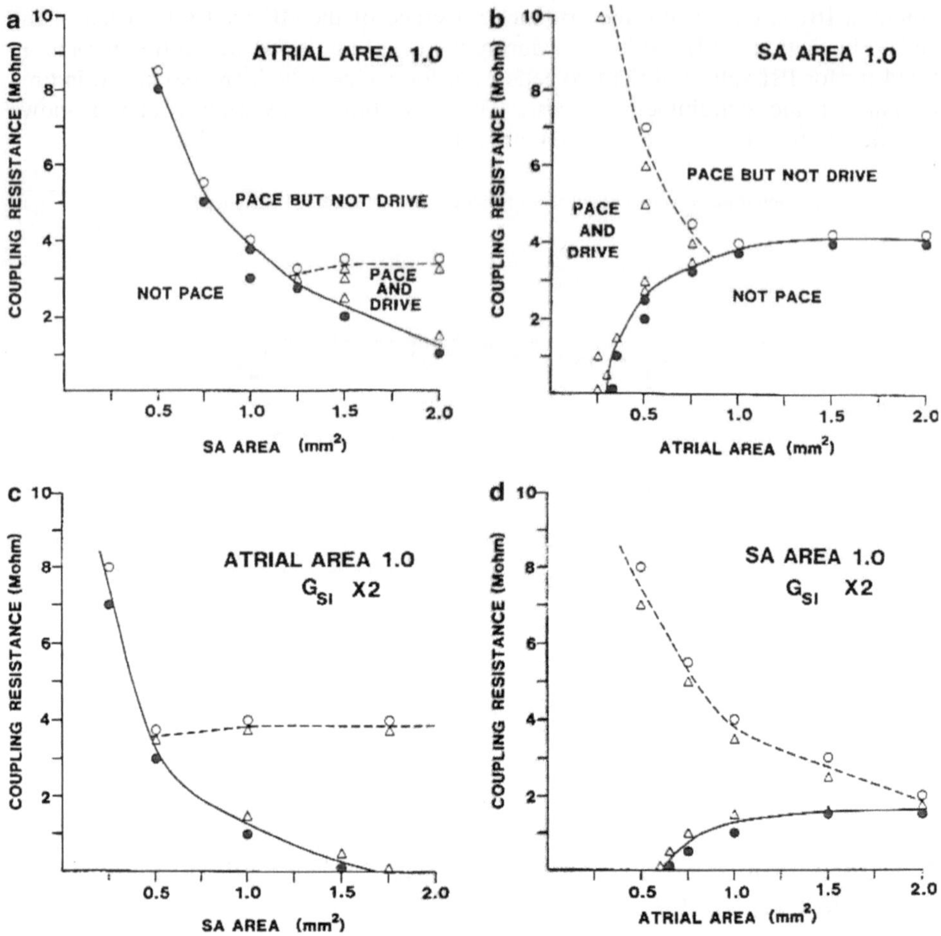

Fig. 4. Simulated results for two electrically coupled membrane regions (as in Fig. 1) for which the SA area is represented by the Irisawa–Noma [10] model and the atrial area is represented by a modified Beeler–Reuter [2] model with shortened action potential duration. Each symbol plotted represents a simulation of 2 s, with the simulation results being either (1) failure of the SA region to pace (NOT PACE, *solid circle*), (2) pacing of the SA region without activation of the atrial region (PACE BUT NOT DRIVE, *open circle*), or (3) pacing of the SA region with subsequent activation of the atrial region (PACE AND DRIVE, *open triangle*). In part **a**, the membrane area of the atrial region. is fixed at 1.0 mm^2 while the membrane area of the SA region and the coupling resistance are varied. In **b**, the membrane area of the SA region is fixed at 1.0 mm^2 while the membrane area of the atrial region and the coupling resistance are varied. The *solid line* separates the graph space into regions of pacing and not pacing. The *dashed line* separates the graph space into regions of successful or unsuccessful driving of the atrial region. **c** and **d** Correspond to **a** and **b**, respectively, except that the magnitude of the slow inward current is doubled for the SA region in c and d. From Joyner et al. [14].

DRIVE, *open circle)*, or (c) pacing of the SA region with subsequent activation of the atrial region (PACE AND DRIVE, *open triangle*). We have not plotted symbols for all of the simulations performed, but only those, which delineate the boundaries between the defined regions on the graph. The solid line separates the graph space into regions of pacing and not pacing. The dashed line separates the graph space into regions of successful or unsuccessful driving of the atrial region. In part a, the membrane area of the atrial region is fixed at 1.0 mm^2 while the membrane area of the SA region and the coupling resistance are varied. Note that, for a fixed SA area of 1 mm^2 (same as for the ATRIAL area of this simulation) there is no value of coupling resistance, which allows successful PACE AND DRIVE. For larger values of SA area (e.g., 1.5 mm^2), the ability to PACE AND DRIVE is limited by the coupling resistance—too low coupling resistance and the SA region does not pace and too high coupling resistance and the SA region can pace but not drive the ATRIAL region. In part b, the membrane area of the SA region is fixed at 1.0 mm^2 while the membrane area of the atrial region and the coupling resistance are varied. The same phenomena are shown as for part a. Part c and d correspond to parts a and b, respectively, except that the magnitude of the slow inward current is doubled for the SA region in parts c and d. The major effect of this change in the properties of the SA region is that the parameter space which now allows successful PACE AND DRIVE is much larger. The boundary between PACE and NOT PACE is shifted to smaller SA area because the diastolic current is enhanced by increasing slow inward current. The boundary between PACE AND DRIVE and PACE BUT NOT DRIVE is also raised to higher values of coupling resistance because of the somewhat elevated peak amplitude of the SA region action potentials.

When we extended these simulations to a radial model of a distributed SAN-Atrial system, we were able to incorporate the differences in the gap junctional coupling among SAN cells as compared to atrial cells [18] with a radially distributed SAN system to investigate the effects of SAN size on the ability to propagate out from the SAN into the surrounding atrial tissue. We used a resistivity of 6,000 Ω cm within the SAN and only 600 Ω cm for the surrounding atrial tissue. This is depicted in the radial diagram in which high coupling resistance is shown as a dark area and low coupling resistance is shown as a gray area. However, we also varied the spatial abruptness with which the two regions merged. For the diagram of part a, we used an abrupt transition at a radial distance of 3 mm from the SAN membrane properties to the atrial membrane properties and the coupling resistance, as shown in the radial diagram and the plot of coupling resistance versus radial distance. For this radial model, the central SAN region paces at a high rate (being electrically somewhat insulated by the high coupling resistivity) but no propagation out into the surrounding atrial tissue occurs, even with an SAN radial distance of 3 mm (see part b of Fig. 5). This failure of propagation was strongly related to the increased electrical load as the action potential approached the SAN-Atrial boundary. When we made this boundary a more gradual transition, as shown in Fig. 5c, we got successful pacing of the SAN as well as propagation out into the atrial area even with an SAN radius of only 1 mm. Note that for the diagram of Fig. 5c and the simulation results of Fig. 5d, the concentric elements with SAN membrane properties occur only within the central 1 mm and the coupling resistance is linearly decreased over the radial distance from 1 to 2 mm, as illustrated in the radial diagram. This suggests that a spatially inhomogeneous distribution of cell types and coupling resistance may be an essential feature of a successful SAN-Atrial interface.

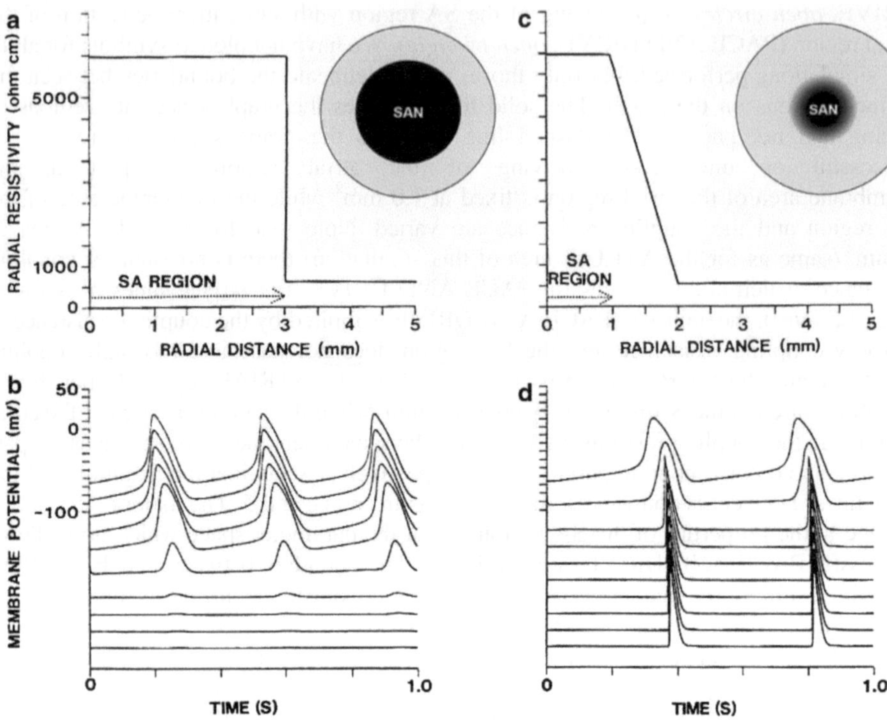

Fig. 5. Results of two simulations with a radial disk model in which the radial resistivity within the SA node region is 6,000 Ω cm. **a** and **c** are diagrams of two different radial distributions of radial resistivity and the SA node model in which diagram **a** has an abrupt decrease in radial resistivity and diagram **c** has a gradual decrease in radial resistivity into the surrounding atrial model cells. **b** and **d** are the simulation results for the distributions **a** and **c**, respectively, with **b** showing the SA node region continuing to PACE BUT NOT DRIVE while **d** shows slowing of the SA node rate but successful PACE AND DRIVE. From Joyner et al. [14].

4 Interactions Among a Spontaneously Pacing Cell Model and a Single Quiescent Real Atrial Cell

With our coupling clamp circuit, we were able to extend these simulations to couple together a model SAN cell, using the Wilders et al. [23] SAN cell model to real isolated rabbit atrial cells. The upper panel of Fig. 6a illustrates action potentials recorded from an isolated uncoupled rabbit atrial cell (solid line) paced by repetitive current pulses of 2 ms duration at a CL of 600 ms. The dashed line shows the steady state solution of membrane potential for the SAN model cell when the model cell is uncoupled (G_C = 0) from the real cell. We established coupling at a variety of conductance values to investigate the interactions between the two cells. The lower panel shows the membrane potential of the atrial cell (solid line) and the membrane

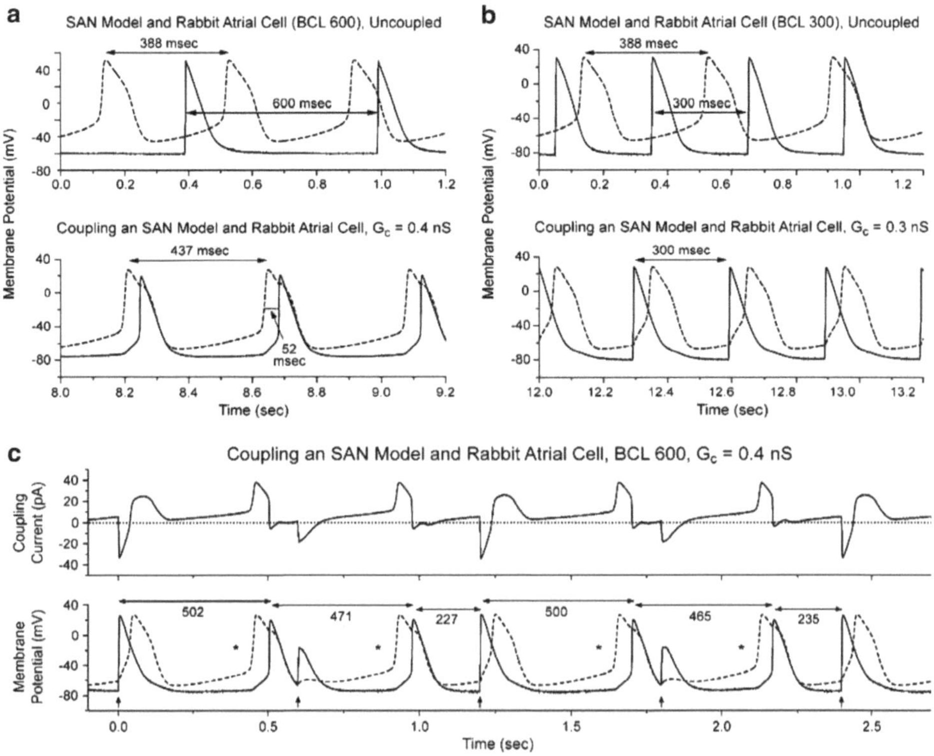

Fig. 6. Interactions between an SAN model cell [22] and a real rabbit atrial cell. **a** The *upper panel* shows action potentials recorded from an isolated uncoupled rabbit atrial cell (*solid line*) paced by repetitive current pulses of 2 ms duration at a basic cycle length (BCL) of 600 ms. The *dashed line* shows the steady state solution of membrane potential for the SAN model cell when the model cell is uncoupled (G_C = 0) from the real cell. The lower panel shows the membrane potential of the atrial cell (*solid line*) and the membrane potential of the SAN model cell (*dashed line*) for G_C = 0.4 nS and no stimuli applied to the atrial cell. **b** The same atrial cell with directly applied stimuli at BCL 300 ms with G_C = 0 nS for the top panel and G_C = 0.3 nS in the lower panel. **c** Same real atrial cell paced at 600 ms with G_C = 0.4 nS coupling to the SAN model. From Joyner et al. [12].

potential of the SAN model cell (dashed line) for G_C = 0.4 nS and no stimuli applied to the atrial cell.

The coupled hybrid cell pair now has an increased CL of 437 ms (indicated by the horizontal arrow in the lower panel), with each AP produced in the SAN model cell accompanied, after a 52 ms delay (arrow), by a driven AP in the real atrial cell. We investigated the effects of a range of G_C for this hybrid cell pair, finding that for values $\geq$ 0.4 nS there was propagation from the SAN model cell to this atrial cell with a decreasing conduction delay as G_C was increased. For G_C < 0.3 nS there was continued pacing of the SAN model without driving of the coupled atrial cell. For 0.3 < G_C < 0.4 nS there was partial synchronization such that only some of the APs from the SAN model cell were conducted to the atrial cell. In Fig. 6b we show the same atrial cell with directly applied stimuli at CL 300 ms with $G_C = 0$ nS for the top panel and $G_C = 0.3$ nS in the lower panel. At this shorter CL, the directly paced atrial cell "overdrives" the SAN model cell with 1:1 conduction from the atrial cell to the SAN model cell.

However, when the directly paced CL of the atrial cell is longer than the intrinsic SAN model CL, arrhythmias may develop. Figure 6c shows results for the same real atrial cell as for Fig. 6a, b when we directly paced the real atrial cell at CL 600 ms and used $G_C = 0.4$ nS. Under these conditions, if a spontaneous activation of the SAN model cell occurs when the real atrial cell is not refractory, propagation from the focus model cell to the real atrial cell may occur. The data shown in Fig. 6c are from a longer recording after a steady state condition had been established. The stimuli to the real atrial cell are shown as vertical arrows in the lower panel. The upper panel shows the coupling current, with a positive polarity indicating current flow from the SAN model cell to the real atrial cell. At time zero there is a stimulus, which directly activates the real atrial cell and this action potential propagates to the SAN model cell. The SAN model cell then has a spontaneous depolarization, which leads to an AP in the SAN model cell (indicated by an asterisk), which then propagates to the real atrial cell. The second direct stimulus to the real atrial cell (at time 0.6 s) now does not activate the real atrial cell because it is refractory. There is then another AP spontaneously generated in the SAN model cell (second asterisk), which also propagates to the real atrial cell. The third direct stimulus to the real atrial cell (at time 1.2 s) does activate the real atrial cell and this AP propagates to the SAN model cell. This process then almost exactly repeats for the next three direct stimuli to the real atrial cell. The APs which occur in the real atrial cell are thus in an arrhythmic pattern, with a repeating series of 3 cycles (~500, 470, and 230 ms for an average CL of 400 ms) for each pair of direct stimuli at CL 600 ms. Note that in this simple two-cell system there is a bidirectional propagation, with the intrinsic automaticity of the focus cell being modulated by the propagation from the directly stimulated cell and also able to propagate to the quiescent cell.

5 Interactions Between a Focus Region and Surrounding Two-Dimensional Tissue

We then used our coupling clamp system to couple together a real spontaneously active nodal cell (isolated from rabbit AV node) to a two-dimensional sheet of model cells, which would represent either atrial or ventricular tissue by using specific models of each tissue type. Since we wanted to use arrays of model cells as surrogates for the electrical characteristics of two-dimensional arrays of either atrial or ventricular real cells, we tested the properties of the two models (for atrial cells [6], for ventricular cells [17]) as to their ability to recreate experimentally recorded differences in excitability of atrial and ventricular cells. For astimulus frequency of 1 Hz, the atrial and ventricular cell models we used produce characteristically different action potential shapes, as expected. The ventricular cell model has a resting potential of –86 mV compared to –80 mV for the atrial cell model. The maximum dV/dt of the ventricular and atrial cell models are 379 and 220 V/s, respectively, with the amplitudes being 136 and 105 mV, respectively. These values are comparable to our previous experimental data [9] for the resting membrane potential, amplitude and maximum d*V*/d*t* measured from ten isolated rabbit atrial cells (–80 ± 1 mV, 109 ± 3 mV, and 206 ± 17 V/s, respectively) which differed significantly ($p < 0.05$) from those values we measured from six isolated rabbit ventricular cells (–82.7 ± 0.4 mV, 127 ± 1.12 mV, and 395 ± 21 V/s, respectively). Several fundamental differences in the membrane currents are included in these models. These differences include a lower value of maximum sodium conductance, lower value of the inward rectifier current (I_{K1}), and greater transient outward current for the atrial model compared to the ventricular model and these differences are based on experimental data as described in the model papers.

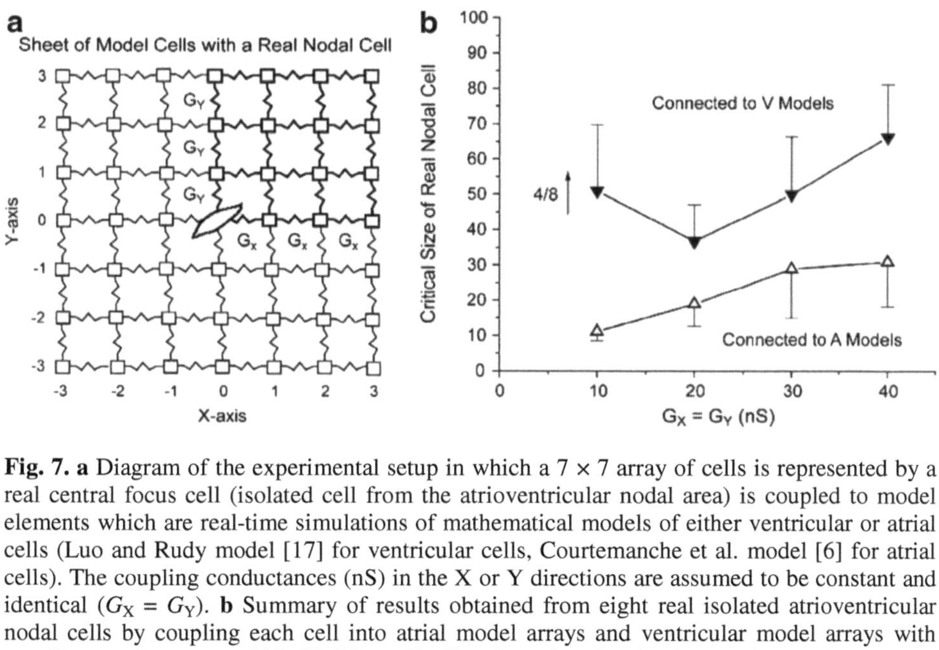

Fig. 7. a Diagram of the experimental setup in which a 7 × 7 array of cells is represented by a real central focus cell (isolated cell from the atrioventricular nodal area) is coupled to model elements which are real-time simulations of mathematical models of either ventricular or atrial cells (Luo and Rudy model [17] for ventricular cells, Courtemanche et al. model [6] for atrial cells). The coupling conductances (nS) in the X or Y directions are assumed to be constant and identical ($G_X = G_Y$). **b** Summary of results obtained from eight real isolated atrioventricular nodal cells by coupling each cell into atrial model arrays and ventricular model arrays with coupling conductances of 10, 20, 30, or 40 nS and varying the size factor of the real nodal cell to determine the critical size which allows spontaneous activity of the nodal cell with propagation into each model array with each value of coupling conductance. For $G_X = G_Y = 10$ nS only 4 of the 8 nodal cells tested were able to successfully propagate into the V model array at any of the sizes (up to 100) tested. Data are shown as mean ± SEM for the V model arrays (*solid triangles*) and the A model arrays (*open triangles*). From Joyner et al. [15].

For these studies, we isolated spontaneously active myocytes from the rabbit atrioventricular node region and then coupled a real nodal cell into a real-time simulation of a small sheet of cells with specified electrical coupling in the *X* and *Y* directions, as shown in Fig. 7a. From eight spontaneously active cells, the average CL when uncoupled was 340 ± 52 ms, with a maximum diastolic depolarization of –65 ± 8 mV, a peak positive amplitude of 36 ± 9 mV, and a maximum d*V*/d*t* of the rising phase of 13.6 ± 3.9 V/s. In the experimental protocol, we allowed the real nodal cell to automatically generate action potentials uncoupled from the sheet for the first 2 s of recording and then we coupled the sheet to the real nodal sheet for a further 10 s of recording. Figure 7b summarizes the results obtained by coupling the same nodal cells into either an atrial model sheet or a ventricular model sheet. The two variables that we used were the coupling conductance between cells of the sheet (G_X and G_Y) and also the ''size'' of the nodal cell. In order to change the effective ''size'' of the real cell, we scaled the current that was injected into the cell to accomplish the coupling to the sheet. Decreasing this current by a factor of two makes the nodal cell effectively two times larger. Generally, decreasing this current by a factor of Z makes the nodal cell effectively Z times larger, as if it consisted of a group of Z cells each well coupled to each other. Figure 7b shows that for a given value of $G_X = G_Y$, the critical size of the nodal region needs to be significantly larger for activation of a ventricular sheet than for an atrial sheet. In addition, the size of the nodal region is required to be larger when $G_X = G_Y$ is increased. Both of these effects can be explained by increases in the electrical load imposed by the sheet on the nodal cells. At the lowest value of $G_X = G_Y$ for the sheet, conduction into the atrial sheet continued but for the ventricular sheet 4 of the 8 cells had conduction block into the ventricular sheet.

6 Influences of Two-Dimensional Tissue Anisotropy and Non-uniform Anisotropy on the Ability of a Focus Region to Propagate

We also did experiments in which we coupled a real excitable cell into a two-dimensional sheet with anisotropy of the coupling conductances [21]. For these experiments we used a real guinea pig ventricular cell as a directly stimulated ''focus'' region to determine if propagation into a simulated sheet of ventricular model cells depended on the anisotropy of G_X and G_Y. The results are summarized in Fig. 8a. For this figure, the open symbols indicate failure of propagation into the sheet while the filled symbols indicate successful propagation into the sheet. For this example, with a size factor of five for the focus cell, propagation failed for $G_X = G_Y$ for values from 10 to 40 nS. However, when the sheet was made anisotropic (e.g., for G_X =30 nS and G_Y either raised to 40 nS or lowered to 20 nS) propagation succeeded. The explanation is clearly not simply the total input resistance as seen from the focus region, but rather that, when the sheet is anisotropic, conduction occurs initially in the direction of the higher coupling conductance and this effectively raises the size of the focus region such that conduction can now occur in the direction of the lower coupling conductance.

A related phenomenon occurs when the sheet has non-uniform anisotropy, as exemplified in the diagram of Fig. 8b. In this scenario we incorporated resistive barriers representing connective tissue separations of cells such that the coupling resistance at that location is infinitively large. Figure 8c, d illustrates the results when the real cell is stimulated with $G_X = G_Y$ =30 nS with no resistive barriers and propagation fails into the sheet. For the same real cell, when we coupled this cell into the sheet with the resistive barriers, propagation succeeded parallel to the barriers and

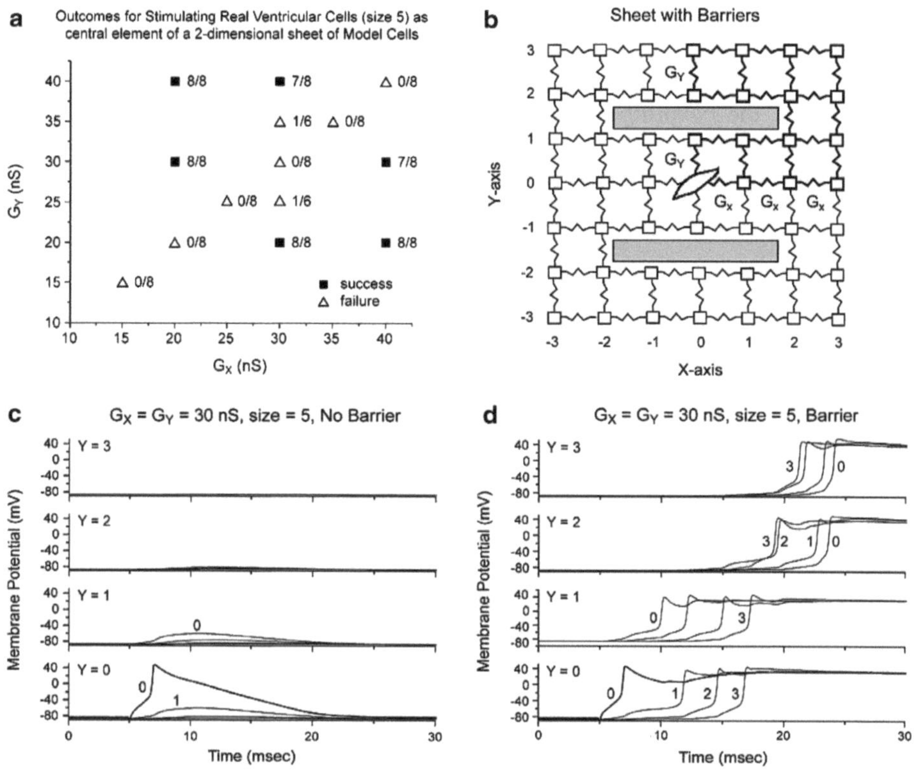

Fig. 8. a Diagram showing success or failure of propagation into the sheet for six real guinea pig ventricular cells as the central element of the sheet diagrammed in Fig. 7a. For each cell, using a size factor of five, we systematically varied G_X and G_Y to test the success or failure of propagation through the sheet when the real cell was repetitively stimulated at 1 Hz. The *filled squares* represent combinations of G_X and G_Y for which propagation was successful and the open symbols represent combinations of G_X and G_Y for which propagation failed. The fractions beside each symbol indicate the ratio of cells with successful propagations to the total number of cells tested with that combination of G_X and G_Y **b** Diagram of a two-dimensional sheet of excitable cells with resistive barriers. Each coupling conductance in the *X* direction has value G_X (nS) and each coupling conductance in the *Y* direction has value G_Y (nS). The central element is represented by a real cell with a variable size and all other elements are represented by real-time simulations of the ventricular membrane model of Luo and Rudy [17] with the additional inclusion of two resistive barriers for which G_Y =0(*gray bars*). **c** Results for the stimulation of a real cell with size factor of five incorporated into the two-dimensional sheet of Fig. 7a (no barriers) with $G_X = G_Y = 30$ nS. The four panels show results for the four rows of elements ($Y = 0$–3) and within each panel the results for each of the four elements ($X = 0$–3) are labeled. The real cell has coordinates (0,0) and is plotted as a thicker line. **d** Results for the same cell and coupling conductances except for the inclusion of the resistive barriers, allowing successful propagation. From Wang et al. [21].

then progressed around the barriers to active the entire sheet (Fig. 8d). Thus, the ability of a focus region to propagate out into surrounding tissue depends not only on the cell type of the tissue and the coupling conductances, but also the spatial distribution of the coupling conductances of the sheet. We also did simulations of propagation from a model SAN region into such spatially inhomogeneously coupled sheets and demonstrated these phenomena for a spontaneously active focus [25].

7 Summary

We have worked over several decades now on determining the design principles for successful propagation from a spontaneously active focus region to a surrounding syncytium of quiescent but excitable myocardium. These principles are of fundamental importance in the functioning of a normal SAN-Atrium interface as well as understanding how an abnormal focus can activate the entire heart. These principles can be briefly summarized as (1) central relative uncoupling to protect the spontaneously firing cells from too much electrotonic inhibition, (2) a transitional region in which the cell type and electrical coupling change from the central SAN region to the peripheral atrial region, and (3) a distributed anisotropy to facilitate focal activity. There is increasing detail from three-dimensional reconstructions of how cell type and connexion type and density are distributed in such regions, such as the elegant studies of Dobrynski et al. [7] Earlier work by Kodama and Boyett [16] demonstrated heterogeneity of action potential types in central versus peripheral regions of the SA node and more recent studies have demonstrated regional differences in gap junction type and distribution [5]. Anatomical studies of the rabbit SAN region show a very disorganized ''mesh'' of cells arranged around ''islands'' of connective tissue. There is also a spatially heterogeneous mixture of typical ''central'' SAN cells and ''atrial'' cells at the peripheral region of the SAN where these cell types can be distinguished by gap junction type as well as by several enzyme markers [7]. This anatomical structure and heterogeneity may provide the central uncoupling, the gradual transition, and the distributed anisotropy, which allows the SAN region to propagate into the atrial tissue. Recent work by Benson et al. [3] demonstrating a specific sodium channel mutation related to sick sinus syndrome in humans may also indicate clinical significance of loading effects on human SAN atrial conduction. The bidirectional effects at the SAN-Atrial interface or at the interface between an ectopic focus in the atrium or the ventricle continue to be of considerable research interest and clinical significance.

References

1. Auricchio A, Klein H (2000) Arrhythmias in heart failure. Curr Treat Options Cardiovasc Med 2(4):329–339
2. Beeler GW, Reuter H (1977) Reconstruction of the action potential of ventricular myocardial fibres. J Physiol 268:177–210
3. Benson DW, Wang DW, Dyment M, Knilans TK, Fish FA, Strieper MJ, Rhodes TH, George AL Jr (2003) Congenital sick sinus syndrome caused by recessive mutations in the cardiac sodium channel gene (SCN5A). J Clin Invest 112(7):1019–1028

4. Cai D, Winslow RL, Noble D (1994) Effects of gap junction conductance on dynamics of sinoatrial node cells: two-cell and large-scale network models. IEEE Trans Biomed Eng 41(3):217–231
5. Coppen SR, Kodama I, Boyett MR, Dobrzynski H, Takagishi Y, Honjo H, Yeh HI, Severs NJ (1999) Connexion45, a major connexion of the rabbit sinoatrial node, is co-expressed with connexion43 in a restricted zone at the nodalcrista terminalis border. J Histochem Cytochem 47(7):907–918
6. CourtemancheM, Ramirez RJ, Nattel S (1998) Ionic mechanisms underlying human atrial action potential properties: insights from a mathematical model. Am J Physiol 275(1 Pt 2):H301–H321
7. Dobrzynski H, Li J, Tellez J, Greener ID, Nikolski VP, Wright SE, Parson SH, Jones SA, Lancaster MK, Yamamoto M, Honjo H, Takagishi Y, Kodama I, Efimov IR, Billeter R, Boyett MR (2005) Computer three-dimensional reconstruction of the sinoatrial node. Circulation 111(7):846–854
8. Efimov IR, Nikolski VP, Rothenberg F, Greener ID, Li J, Dobrzynski H, Boyett M (2004) Structure function relationship in the AV junction. Anat Rec A Discov Mol Cell Evol Biol 280(2):952–965
9. Golod DA, Kumar R, Joyner RW (1998) Determinants of action potential initiation in isolated rabbit atrial and ventricular myocytes. Am J Physiol 274:H1902–H1913
10. Irisawa H, Noma A (1982) Pacemaker mechanisms of rabbit sinoatrial node cells. In: Bouman HN, Jongsma HJ (eds) Cardiac rate and rhythm. Martinus Nihhoff Publishers, The Hague, pp 35–52
11. Janse MJ, Opthof T, Kleber AG (1998) Animal models of cardiac arrhythmias. (review) (163 refs). Cardiovasc Res 39(1):165–177
12. Joyner RW, Kumar R, Golod DA, Wilders R, Jongsma HJ, Verheijck EE, Bouman L, Goolsby WN, van Ginneken AC (1998) Electrical interactions between a rabbit atrial cell and a nodal cell model. Am J Physiol 274(6 Pt 2):H2152–H2162
13. Joyner RW, Picone J, Veenstra R, Rawling D (1983) Propagation through electrically coupled cells. Effects of regional changes in membrane properties. Circ Res 53(4):526–534
14. Joyner RW, van Capelle FJL (1986) Propagation through electrically coupled cells: how a small SA node drives a large atrium? Biophys J 50:1157–1164
15. Joyner RW, Wang YG, Wilders R, Golod DA, Wagner MB, Kumar R, Goolsby WN (2000) A spontaneously active focus drives a model atrial sheet more easily than a model ventricular sheet. Am J Physiol Heart Circ Physiol 279(2):H752– H763
16. Kodama I, Boyett MR (1985) Regional differences in the electrical activity of the rabbit sinus node. Pflugers Arch 404:214–226
17. Luo CH, Rudy Y (1994) A dynamic model of the cardiac ventricular action potential. I. Simulations of ionic currents and concentration changes. Circ Res 74:1071–1096
18. Masson Pevet MA, Bleeker WK, Besselsen E, Mackaay AJC, Jongsma HJ, Bouman LN (1982) On the ultrastructural identification of pacemaker cell types. In: Bouman LN, Jongsma HJ (eds) Cardiac rate and rhythym. Martinus Nijhoff, The Hague, pp19–34
19. Veenstra RD, DeHaan RL (1986) Electrotonic interactions between aggregates of chick embryo cardiac pacemaker cells. Am J Physiol 250:H453–H463
20. Verheijck EE, Wilders R, Joyner RW, Golod DA, Kumar R, Jongsma HJ, Bouman LN, van Ginneken AC (1998) Pacemaker synchronization of electrically coupled rabbit sinoatrial node cells. J Gen Physiol 111:95–112

21. Wang YG, Kumar R, Wagner MB, Wilders R, Golod DA, Goolsby WN, Joyner RW (2000) Electrical interactions between a real ventricular cell and an anisotropic two-dimensional sheet of model cells. Am J Physiol Heart Circ Physiol 278(2):H452–H460
22. Wilders R, Jongsma HJ (1993) Beating irregularity of single pacemaker cells isolated from the rabbit sinoatrial node. Biophys J 65:2601–2613
23. Wilders R, Jongsma HJ, van Ginneken AC (1991) Pacemaker activity of the rabbit sinoatrial node. A comparison of mathematical models. Biophys J 60:1202–1216
24. Wilders R, Kumar R, Joyner RW, Jongsma HJ, Verheijck EE, Golod D, van Ginneken AC, Goolsby WN (1996) Action potential conduction between a ventricular cell model and an isolated ventricular cell. Biophys J 70(1):281–295
25. Wilders R, Wagner MB, Golod DA, Kumar R, Wang YG, Goolsby WN, Joyner RW, Jongsma HJ (2000) Effects of anisotropy on the development of cardiac arrhythmias associated with focal activity. Pflugers Arch 441(2–3):301– 312

Computer Modelling of the Sinoatrial Node

Ronald Wilders

Department of Physiology, Academic Medical Center,
University of Amsterdam, Meibergdreef 15,
1105 AZ Amsterdam, The Netherlands
r.wilders@amc.uva.nl

Abstract. Over the past decades patch-clamp experiments have provided us with detailed information on the different types of ion channels that are present in the cardiac cell membrane. Sophisticated cardiac cell models based on these data can help us understand how the different types of ion channels act together to produce the cardiac action potential. In the field of biological pacemaker engineering, such models provide important instruments for the assessment of the functional implications of changes in density of specific ion channels aimed at producing stable pacemaker activity. In this review, an overview is given of the progress made in cardiac cell modelling, with particular emphasis on the development of sinoatrial (SA) nodal cell models. Also, attention is given to the increasing number of publicly available tools for non-experts in computer modelling to run cardiac cell models.

Keywords: Sinoatrial node, Computer simulations, Action potentials, Electrophysiology, Heart.

1 Introduction

Biological pacemakers have been obtained by suppression of current encoded by the Kir2 gene family (inward rectifier potassium current, I_{K1}) or overexpression of current encoded by the HCN gene family (hyperpolarization-activated current, I_f) see review by Rosen et al. ([74], Chapter 5). The development of biopacemakers has raised the interest in mechanisms of pacemaker generation and means to stabilize and control pacemaker activity, as illustrated by the recent paper by de Boer et al. ([9], Chapter 7). Like the experimental approaches to create a biological pacemaker, computer simulation studies have focused on the effects of I_{K1} downregulation or I_f upregulation in ventricular myocytes [45, 79, 86]. These computer simulations demonstrated that the mechanism of pacemaker generation in I_{K1}-downregulated ventricular myocytes relies on the net current generated by the sodium–calcium exchanger (I_{NaCa}) as the carrier of pacemaker current causing phase-4 depolarization [45, 79] and that I_f upregulation may be associated with cycle length instability [86].

In contrast to the biopacemakers engineered so far, sinoatrial (SA) nodal pacemaker cells are characterized by the expression of several depolarizing currents that are all involved in highly stable phase-4 depolarization and pacemaking. These include background sodium current ($I_{b,Na}$), L-and T-type calcium current ($I_{Ca,L}$ and

J.A.E. Spaan et al. (Eds.): Biopacemaking, BIOMED, pp. 121–148, 2007.
springerlink.com

$I_{Ca,T}$, respectively), I_f, I_{NaCa}, and a sustained inward current (I_{st}). Furthermore, the acetylcholine-sensitive outward potassium current ($I_{K,ACh}$) provides vagal control of pacemaker rate. Each of these currents provides a potential target for pacemaker regulation. Further insight into pacemaker mechanisms and control of pacemaker activity may be obtained through the use of computer models of electrical activity of SA nodal cells. The aim of the present review is to provide an overview of the available single cardiac cell models, with particular emphasis on single SA nodal cell models, and the tools that are available for non-experts in computer modelling to run these models. This review is thus intentionally restricted to single cardiac cell models. Propagation of pacemaker activity is addressed elsewhere in this issue in a review by Joyner et al. ([40], Chapter 8).

2 Development of Cardiac Cell Models

Back in 1928, van der Pol and van der Mark [84] presented the first mathematical description of the heartbeat, which is in terms of a relaxation oscillator. Their work has given rise to a family of models of nerve and heart 'cells' (excitable elements) in terms of their key properties, i.e. excitability, stimulus threshold, and refractoriness. These models are relatively simple with a minimum number of equations and variables [23, 83]. Because they are compact, these models have been widely used in studies of the spread of excitation in tissue models consisting of large numbers of interconnected 'cells' (e.g. [83]). A major drawback of this family of models is the absence of explicit links between the electrical activity and the underlying physiological processes like the openings and closures of specific ion channels. Today's sophisticated cardiac cell models provide such links and are all built on the framework defined by the seminal work of Hodgkin and Huxley [34], for which they received the 'Nobel Prize in Physiology or Medicine' in 1963.

2.1 The Seminal Hodgkin–Huxley Model

Hodgkin and Huxley investigated the electrical activity of the squid giant axon, on which they published a series of five (now classical) papers in 1952. In their concluding paper, they summarized their experimental findings and presented "a quantitative description of membrane current and its application to conduction and excitation in nerve" [34]. This "quantitative description" included a mathematical model derived from an electrical equivalent of the nerve cell membrane. They identified sodium and potassium currents flowing across the giant axon membrane and represented these in terms of the sum of conductive components, what we now identify as ion channels, and membrane capacitance. Figure 1 shows the associated electrical circuit diagram in its simplest form. The membrane current (I_m) is a function of the voltage across the cell membrane (V_m), the equilibrium potential (E_m) of the ions carrying this membrane current, and the membrane conductance (g_m). Because no current is entering or leaving the electrical circuit, the sum of the capacitative current and I_m equals zero. As a consequence (Fig. 1), we have

$$C_m \times dV_m/dt = -I_m \quad (1)$$

which is the differential equation relating membrane potential to net membrane current that forms the basis of all cardiac cell models. Hodgkin and Huxley demonstrated that g_m can be separated into sodium and potassium components, which are both functions of voltage and time. In their analysis they introduced the concept of activation and inactivation 'gates' and provided equations governing the time and voltage dependence of these gates. In its simplest form, with a single gate controlling the state of each channel, the ionic current flowing through 'gated' ion channels is given by

$$I_{ion} = x \times g_{max} \times (V_m - E_{ion}), \tag{2}$$

Where g_{max} is the maximal conductance (all channels open), x the proportion of g_{max} actually available, and E_{ion} the reversal potential. The activation 'gating variable' x, i.e. the fraction of active (open) gates (ranging between 0 and 1), changes with time as

$$dx/dt = \alpha \times (1-x) - \beta \times x, \tag{3}$$

where α and β are first-order 'rate constants' that are both functions of membrane potential. The link between theory and voltage-clamp experiment is that the rate constants can be derived from the experimental steady-state activation curve (x_∞) and time constant of activation (τ_x) through

$$x_\infty = \alpha/(\alpha + \beta) \tag{4}$$

and

$$\tau_x = 1/(\alpha + \beta). \tag{5}$$

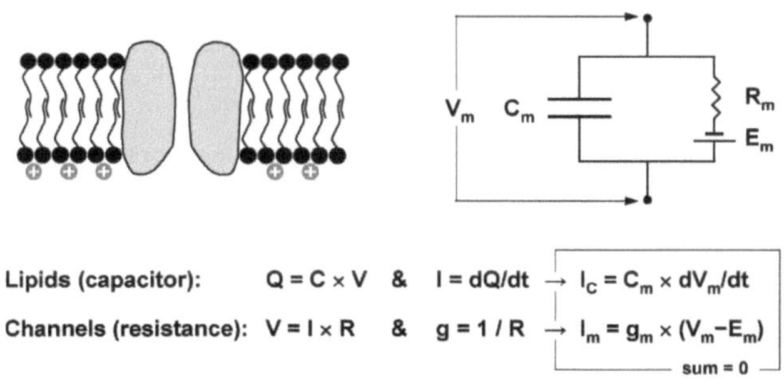

Fig. 1. Electrical equivalent of the cardiac cell membrane. The lipid bilayer acts as a capacitor with capacitance C_m. The current charging the capacitor (I_C) is related to the voltage across the cell membrane (V_m) through $I_C = C_m \times dV_m/dt$. The ion channel in the cell membrane acts as a resistor with resistance R_m, or conductance g_m, which is expressed in 'mho' or 'siemens' (S). The current flowing through the ion channel (I_m) is related to V_m through $I_m = g_m \times (V_m - E_m)$, where E_m is the Nernst or reversal potential for the ions passing through this particular channel.

As illustrated in Fig. 2, today's ionic cardiac cell models still largely follow the concept of Hodgkin and Huxley. The kinetics of ionic currents that have been identified in patch-clamp experiments are described in terms of first-order 'activation' and 'inactivation' gating processes, through Eqs. 4 and 5, and turned into components of an electrical circuit, which underlies the set of model equations. The main extension to the concept of Hodgkin and Huxley is the inclusion of an additional set of equations governing the changes in intracellular ion concentrations (sodium, potassium, calcium) and the intracellular calcium uptake and release processes ('second-generation models'; see below).

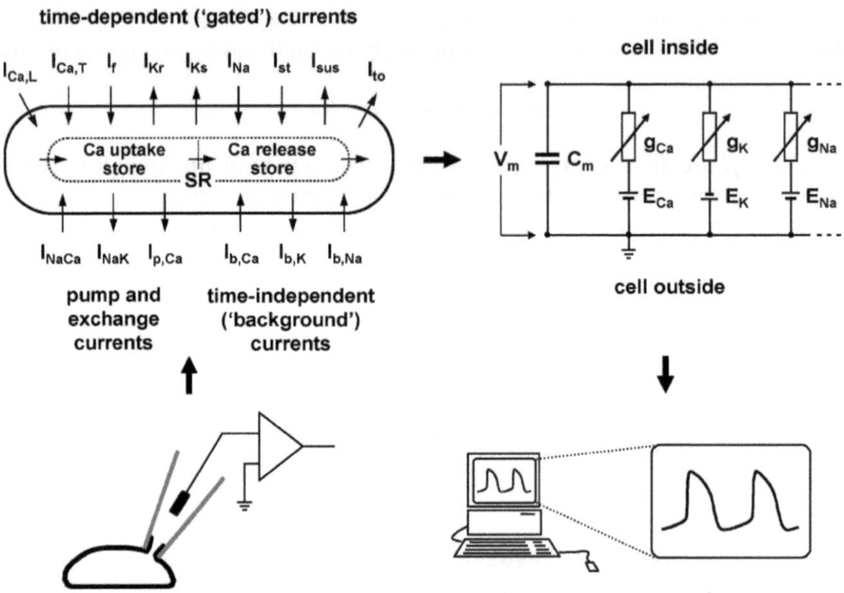

Fig. 2. Common layout of cardiac cell models. Data from patch-clamp experiments (*bottom left*) provide quantitative information on each of the individual membrane current components (*top left*). This information is turned into an electrical equivalent (*top right*; cf. Fig. 1). The resulting equations, together with equations describing intracellular processes, e.g. the calcium uptake and release by the sarcoplasmic reticulum (SR), are then compiled into a computer model of the cardiac cell, an SA nodal cell in this case (*bottom right*). The *upper left diagram* shows all ionic currents that have been incorporated into SA nodal cell models: L-type calcium current ($I_{Ca,L}$), T-type calcium current ($I_{Ca,T}$), hyperpolarization-activated current (I_f), rapid and slow delayed rectifier potassium current (I_{Kr} and I_{Ks}, respectively), fast sodium current (I_{Na}), sustained inward current (I_{st}), sustained and transient components of the 4-aminopyridine-sensitive (4-AP-sensitive) outward current (I_{sus} and I_{to}, respectively), sodium–calcium exchange current (I_{NaCa}), sodium–potassium pump current (I_{NaK}), sarcolemmal calcium pump current ($I_{p,Ca}$), and calcium, potassium, and sodium background current ($I_{b,Ca}$, $I_{b,K}$ and $I_{b,Na}$, respectively).

2.2 Early Cardiac Cell Models (1960–1988)

The Hodgkin–Huxley concept was first applied to cardiac cells by Noble [59, 60]. His model of cardiac Purkinje fibres was relatively simple and had only five variables. Upon availability of more experimental data, the model was updated twice, in the 1970s by McAllister et al. [57] and in the 1980s by DiFrancesco and Noble [14]. The latter two models served as a basis for the development of more specific models of ventricle, atrium, and sinoatrial node (Table 1).

2.3 Detailed Cardiac Cell Models (1989–2006)

According to current standards, the models developed during the first decades of cardiac cell modelling (Table 1) were simple, with a small number of variables and equations, and of generic nature, with limited specificity regarding species or location within the heart. This changed rapidly during the subsequent period. Starting in the late 1980s, a large number of cardiac cell models of increasing complexity have been developed, taking advantage of the immense progress made in cardiac cellular electrophysiology and the powerful computer resources that had become available (Table 2). Among these models, the 'Luo–Rudy II' ventricular cell model [52], also known as 'phase-2 Luo–Rudy' or 'LR2' model, has become a classical one. Several more recent ventricular and atrial cell models have been built upon the Luo–Rudy equations.

Table 1. Early ionic models of mammalian cardiac cells (1960–1988)

Model	Parent model
Purkinje fibre models	
Noble [59, 60]	–
McAllister et al. [57]	Noble [59, 60]
DiFrancesco and Noble [14]	McAllister et al. [57]
Ventricular cell models	
Beeler and Reuter [1]	McAllister et al. [57]
Drouhard and Roberge [18]	Beeler and Reuter [1]
Atrial cell models	
Hilgemann and Noble [33]	DiFrancesco and Noble [14]
Sinoatrial cell models (rabbit)	
Yanagihara et al. [91]	–
Irisawa and Noma [36]	Yanagihara et al. [91]
Bristow and Clark [6]	McAllister et al. [57]
Noble and Noble [61]	DiFrancesco and Noble [14]

Since the Luo–Rudy II model was published in 1994, it has been updated several times, thus creating the 'LRd model'. Although Table 2 may look exhaustive, it is certainly not complete. In particular, many updates or extensions of models listed in Table 2 exist. These can easily be retrieved through PubMed or related electronic tools.

Table 2. Detailed ionic models of mammalian cardiac cells (1989–2006)

Model	Species	Parent model
Ventricular cell models		
Noble et al. [63]	Guinea pig	Earm and Noble [19]
Luo and Rudy [51]	Guinea pig	Beeler and Reuter [1]
Nordin [66]	Guinea pig	–
Luo and Rudy [52]	Guinea pig	Luo and Rudy [51]
Jafri et al. [39]	Guinea pig	Luo and Rudy [52]
Noble et al. [65]	Guinea pig	Noble et al. [63]
Priebe and Beuckelmann [70]	Human	Luo and Rudy [52]
Winslow et al. [90]	Canine	Jafri et al. [39]
Pandit et al. [68]	Rat	Demir et al. [11]
Puglisi and Bers [71]	Rabbit	Luo and Rudy [52]
Bernus et al. [2]	Human	Priebe and Beuckelmann [70]
Fox et al. [24]	Canine	Winslow et al. [90]
Greenstein and Winslow [28]	Canine	Winslow et al. [90]
Cabo and Boyden [7]	Canine	Luo and Rudy [52]
Matsuoka et al. [55]	Guinea pig	–
Bondarenko et al. [3][a]	Mouse	–
Shannon et al. [77][a]	Rabbit	Puglisi and Bers [71]
ten Tusscher et al. [81]	Human	–
Iyer et al. [38][a]	Human	–
Hund and Rudy [35]	Canine	Luo and Rudy [52]
ten Tusscher and Panfilov [80]	Human	ten Tusscher et al. [81]
Atrial cell models		
Earm and Noble [19]	Rabbit	Hilgemann and Noble [33]
Lindblad et al. [48]	Rabbit	–
Courtemanche et al. [8]	Human	Luo and Rudy [52]
Nygren et al. [67]	Human	Lindblad et al. [48]
Ramirez et al. [72]	Canine	Courtemanche et al. [8]
Sinoatrial cell models		
Noble et al. [62]	Rabbit	Noble and Noble [61]
Wilders et al. [89]	Rabbit	Noble and Noble [61]
Demir et al. [11]	Rabbit	–
Dokos et al. [16]	Rabbit	Wilders et al. [89]
Dokos et al. [17]	Rabbit	Dokos et al. [16]
Demir et al. [12]	Rabbit	Demir et al. [11]
Endresen et al. [20]	Rabbit	–
Zhang et al. [93]	Rabbit	–
Boyett et al. [5]	Rabbit	Zhang et al. [93]
Zhang et al. [94]	Rabbit	Zhang et al. [93]
Kurata et al. [43]	Rabbit	–
Sarai et al. [75]	Rabbit	–
Lovell et al. [50][a]	Rabbit	–
Mangoni et al. [54]	Mouse	Zhang et al. [93]

[a] Model has Markov-type channel gating

2.4 First-Generation and Second-Generation Models

In the early ionic models of cardiac cells, all ion concentrations were constant, so that no provision had to be made for pumps and exchangers to regulate these concentrations. We refer to those models as ‘first-generation’ models, in contrast to ‘second-generation’ models with time-varying ion concentrations that are regulated by pumps and exchangers. As set out in detail by Krogh-Madsen et al. [42], there are two major problems with the more physiologically realistic second-generation models, in which, besides membrane potential and gating variables, ion concentrations vary in time. The first is drift, with very slow long-term trends in some of the variables, primarily ion concentrations. Drift has been dealt with in several ways, including stimulus current assignment to a specific ionic species (e.g. [41]). The other major problem noted with second-generation models is ‘degeneracy’. This means that there is a continuum of equilibrium points, i.e. ‘steady-state solutions’ to the set of differential equations defining the model, rather than isolated equilibrium points, e.g. the resting potential of a quiescent system depends on the initial conditions. This can be overcome through a ‘chemical’ approach using an explicit formula for the membrane potential of cells in terms of the intracellular and extracellular ion concentrations [20]. A practical solution to the problems of drift and degeneracy is to set the intracellular sodium and potassium concentrations to constant values, which reflects the buffering of these ion concentrations in patch-clamp experiments through the pipette solution.

When using cardiac cell models, one should realize that first-generation models tend to reach a steady state within only a few beats, whereas second-generation should be run for a much longer time to reach steady state. For example, in the canine atrial cell model by Kneller et al. [41] action potential duration reaches steady state after approximately 40 min of pacing, which limits its suitability for performing large-scale (whole-heart) simulations, where simulating a single beat may already take several hours on a state-of-the-art computer [69]. Furthermore, it should be emphasized that the equations representing the calcium subsystem of the second-generation models are still evolving and even the latest comprehensive models fail to adequately represent fundamental properties of calcium handling and inactivation of L-type calcium current by intracellular Ca^{2+} (see, e.g. [38, 82] and primary references cited therein).

2.5 Deterministic, Stochastic and Markov Models

Until a few years ago, cardiac cell models employed the traditional Hodgkin–Huxley formulation of ion channel gating, with one or more independent gates that can flip between their open and closed state (Fig. 3a), although it had been noted that many currents are more accurately described by Markov-type models. Markov models are not restricted to the Hodgkin–Huxley concept of independent gates defining the channel state diagram. Thus, they allow for more complex state diagrams with specific transitions between open, closed, or inactivated channel states (Fig. 3b), with the advantage of being more closely related to the underlying structure and

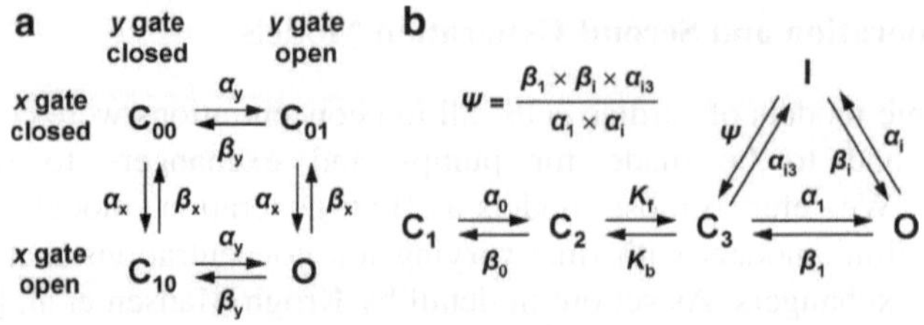

Fig. 3. Hodgkin–Huxley and Markov-type models of I_{Kr} channel gating. **a** State diagram of classical Hodgkin–Huxley-type model. The conductive state of the channel is controlled by two independent activation and inactivation gates (*x* and *y*, respectively), resulting in four different channel states. C_{00}, C_{01}, and C_{10} are the closed states of the channel and O is the open state. **b** State diagram of Markov-type scheme [56]. C_1, C_2, and C_3 are closed states, O is the open state, and I the inactivated state. All transition rates, except K_f and K_b, are a function of membrane potential. Ψ is defined as a function of other transition rates to satisfy the microscopic reversibility condition.

conformation of the ion channel proteins. A disadvantage is that Markov models often add several differential equations (and parameters) to the model, thus increasing the model complexity and computation time. In the recent model of a mouse ventricular cell by Shannon et al. [77], Markov channel models were therefore generally avoided unless they were determined to be necessary.

Markov models should not be confused with stochastic models. Cardiac cells models employing Markov-type channel gating are as deterministic as models employing Hodgkin–Huxley channel gating. In either case, the number of ion channels occupying a certain channel state is represented as a fraction (between 0 and 1) that changes with time as a continuous number. Stochastic models are non-deterministic (probabilistic) models in which the stochastic 'random' openings and closures of individual ion channels are taken into account. The number of ion channels occupying a certain channel state is then represented as a discrete number that changes 'randomly' and in discrete steps with time. As a consequence, action potentials show 'channel noise' and beat-to-beat fluctuations, like experimental recordings [88, 92]. Stochastic ionic models of SA nodal pacemaker cells have been developed by Guevara and Lewis [29] and Wilders et al. [88], whereas Greenstein and Winslow have developed an ionic model of the canine ventricular myocyte incorporating stochastic gating of sarcolemmal L-type calcium channels and ryanodine-sensitive sarcoplasmic reticulum (SR) calcium release channels [28].

2.6 Computational Aspects

The original computations by Hodgkin and Huxley [34] were done by hand. As memorized by Huxley in his Nobel lecture, "this was a laborious business: a membrane action took a matter of days to compute, and a propagated action potential took a matter of weeks." For his initial Purkinje fibre action potential simulations, Noble [59, 60] could make use of the university computer for which he had to write a program in machine code. It took 2 h of CPU time to simulate a single action potential model (with only five variables). In the 1970s, Beeler and Reuter [1] used a mainframe computer to solve the differential equations of their 8-variable ventricular cell model. A single action potential could be computed in about 40 s. In subsequent decades, computational power has grown exponentially. Simulating a single Noble or Beeler– Reuter action potential now takes no more than one or a few milliseconds on a desktop personal computer. Nevertheless, computational efficiency is still an important issue when modelling cardiac cells. Models have not only become much more complex with many more variables, but also require much longer time to reach steady state (see paragraph 2.4). Computational efficiency is, of course, of particular importance in large-scale tissue or whole-heart simulations.

The ongoing demand for computational power is illustrated by the development of human ventricular cell models. The models by Priebe and Beuckelmann [70] and ten Tusscher et al. [81], with 15 and 16 variables, respectively, are of a similar complexity. Both use Hodgkin–Huxley-type equations for the ionic currents and they model intracellular sodium, potassium, and calcium dynamics. The Iyer et al. [38] human ventricular cell model uses Markov-type models rather than Hodgkin–Huxley-type equations to describe the dynamics of the major ionic currents. As set out above, this allows one to fit single channel experimental data and to incorporate knowledge on ion channel structure, but this comes at the cost of a much higher number of variables, as many as 67 in the case of the Iyer et al. model. Apart from the number of variables, the computational efficiency of a model is also determined by the 'stiffness' of its equations: a model with 'stiff' equations requires a small integration time step for a stable and precise solution of its differential equations. ten Tusscher et al. [82] compared the computational efficiency of the above human ventricular cell models under standardized conditions, using simple Euler forward integration, which is a widely used integration method in cardiac cell modelling. They observed that the Priebe–Beuckelmann and ten Tusscher et al. models allowed for a much larger integration time step (20 μs) than the Iyer et al. model (0.02 μs). Together, the differences in the number of variables and the integration time step turn simulations with the Priebe–Beuckelmann and ten Tusscher et al. models to be both relatively fast, whereas the Iyer et al. model is almost 1,000 times 'slower' [82].

2.7 Multicellular Simulations

The 'bidomain' model has been widely accepted as the major approach for theoretical and numerical investigation of macroscopic electric phenomena in cardiac tissue. It is based on the representation of the tissue as two interpenetrating extra- and intracellular domains, each of them having different conductivities along and across the direction of the fibres [32]. The state variables describing the system are the local intracellular and extracellular potentials (ϕ_i and ϕ_e, respectively), with the transmembrane potential defined as $V_m = \phi_i - \phi_e$. Although the bidomain model gives the most accurate approximation of whole heart tissue, it requires considerable computational power for calculations, both in terms of processor speed and computer memory: simulating a single cardiac cycle in a bidomain whole-heart model may take as long as two days on a state-of-the-art 32-processor computer [69]. This explains why large-scale simulations are often 'monodomain', i.e. ϕ_e is ignored and set to zero. It has been shown that, in the absence of applied currents, a monodomain model is sufficient to study propagating action potentials on the scale of a human heart [69]. Ionic models of single cardiac cells are also 'monodomain': as in experiments on isolated cardiac cells, the extracellular space is grounded to earth (cf. Figs. 1, 2). Given the required computational power, only simple first-generation cardiac cell models, with few variables and equations, can be used in large-scale bidomain simulations. Thus, bidomain simulations do not permit highly detailed studies of effects of individual ionic currents on cardiac activation. This also holds, to a lesser extent, for monodomain simulations. Cardiac cell models that are typically used in large-scale simulations include the simplified ionic models by Fenton and Karma [21] and Bernus et al. [2].

3 SA Nodal Cell Models

Starting with the model by Yanagihara et al. [91], a number of SA nodal cells has been developed (Tables 1, 2). The complexity of the models has increased considerably over the years as new currents have been identified and more experimental data have become available. Data have been obtained almost exclusively from experiments on rabbit. Therefore, it is not surprising that, with the exception of the recently developed mouse SA nodal cell model by Mangoni et al. [54], all SA nodal models are in effect models of rabbit SA nodal cells. The Hodgkin–Huxley formalism of channel gating has proven successful in SA nodal cell modelling, with only one model employing Markov-type channel gating [50].

In this section, an overview of 25 years of SA nodal cell modelling is given, illustrated with spontaneous electrical activity generated in 12 different SA nodal cell models that have published since 1980 (Figs. 4, 5, 6; Table 3). Subsets of these models have previously been compared by Wilders et al. [89], Garny et al. [25], and Kurata et al. [43]. In the study by Wilders et al. [89], the models by Bristow and Clark [6], Irisawa and Noble [36], Noble and Noble [61], and Noble et al. [62] were compared, whereas Garny et al. [25] compared the models by Demir et al. [11], Dokos et al. [16], and Zhang et al. [93] as well as an updated version of the Noble et al. [62] model, as incorporated in the commercially available OxSoft Heart 4.X software package (see paragraph 4.2). In their presentation of a novel SA nodal cell model, Kurata et al. [43] have compared this model to the models by Wilders et al. [89], Demir et al. [11], Dokos et al. [16], and Zhang et al. [93].

3.1 SA Nodal Cell Models of the 1980s: Generic Models

Realistic mathematical models of electrical activity of cardiac cells could not be constructed before electrophysiological techniques allowed measurements of individual membrane currents. The McAllister–Noble–Tsien Purkinje fibre model (MNT model [57]) was the first mathematical model of electrical activity of cardiac cells that was based on experimental data on individual membrane currents. Two years later, Beeler and Reuter [1] published their widely used mathematical model of electrical activity of the ventricular myocardial cell, based on the MNT model (Table 1). Several years later, Bristow and Clark [6] presented a mathematical model of (multicellular) SA node pacemaker activity that was constructed by modifying the MNT model. The membrane current components of this Purkinje fibre model were modified to reproduce a particular SA nodal "reference transmembrane potential waveform." The action potential and associated ionic currents generated by this

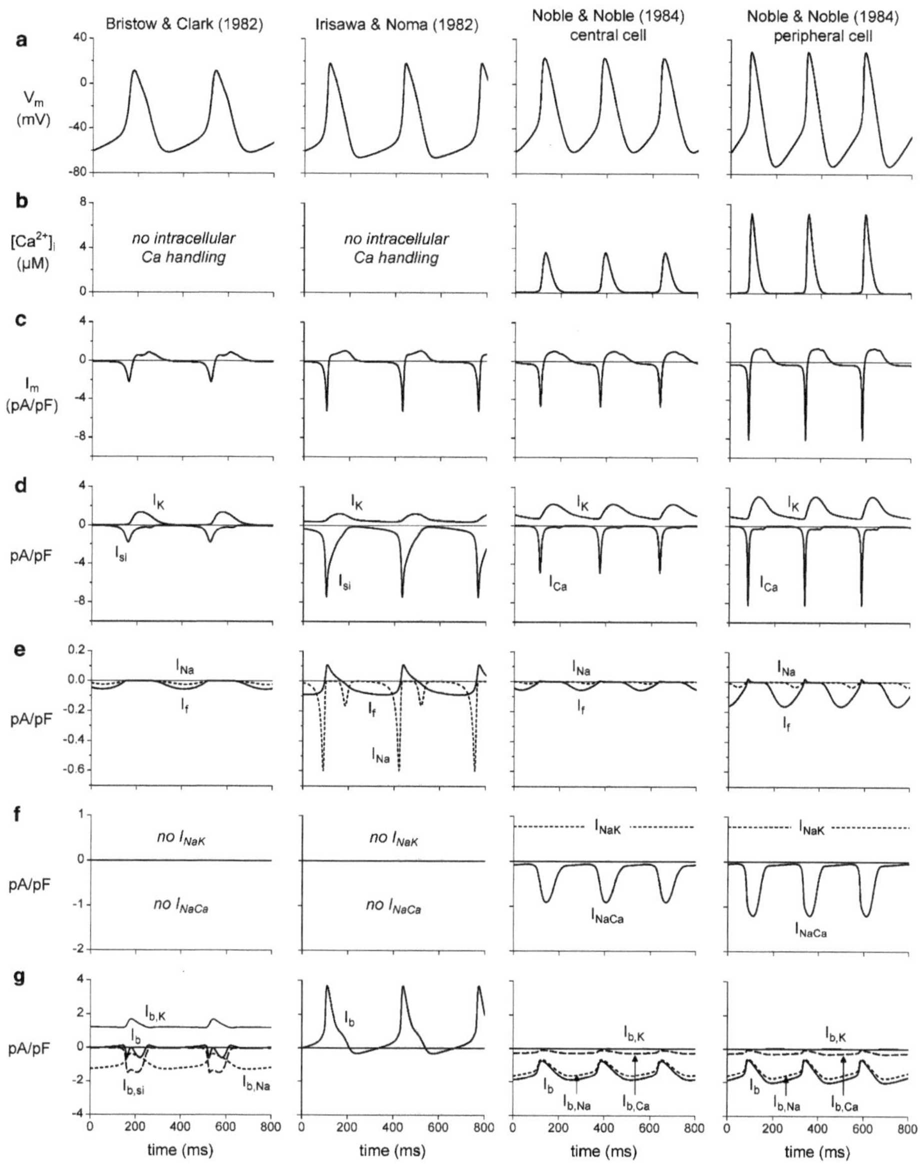

Fig. 4. Action potentials, calcium transients, and transmembrane ionic currents during 800 ms of spontaneous electrical activity in the rabbit SA node models of Bristow and Clark [6], Irisawa and Noma [36], and Noble and Noble [61]. **a** Membrane potential (V_m). **b** Intracellular free calcium concentration ($[Ca^{2+}]_i$). **c** Net transmembrane current (I_m). **d** Delayed rectifier potassium current (I_K) and slow inward or calcium current (I_{si} and I_{Ca}, respectively). **e** Fast sodium current (I_{Na}) and hyperpolarization-activated current (I_f). **f** Sodium–potassium pump current (I_{NaK}) and sodium–calcium exchange current (I_{NaCa}). **g** Net background current (I_b) and its calcium, potassium, and sodium components ($I_{b,si}$ or $I_{b,Ca}$, $I_{b,K}$, and $I_{b,Na}$, respectively).

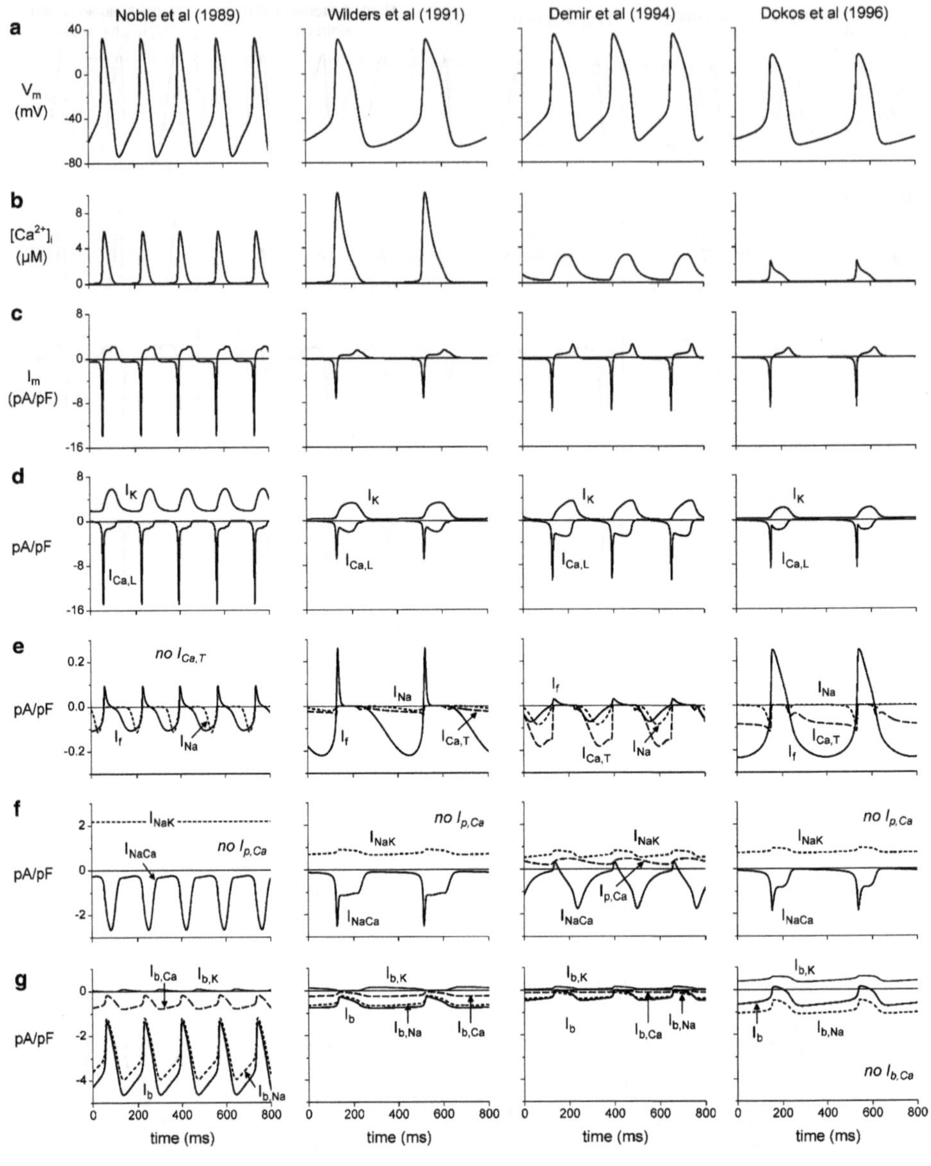

Fig. 5. Action potentials, calcium transients, and transmembrane ionic currents during 800 ms of spontaneous electrical activity in the single rabbit SA nodal cell models of Noble et al. [62], Wilders et al. [89], Demir et al. [11], and Dokos et al. [16]. **a** Membrane potential (V_m). **b** Intracellular free calcium concentration ($[Ca^{2+}]_i$). **c** Net transmembrane current (I_m). **d** Delayed rectifier potassium current (I_K) and L-type calcium current ($I_{Ca,L}$). **e** Fast sodium current (I_{Na}), hyperpolarization-activated current (I_f), and T-type calcium current ($I_{Ca,T}$). **f** Sodium–potassium pump current (I_{NaK}), sodium–calcium exchange current (I_{NaCa}), and sarcolemmal calcium pump current ($I_{p,Ca}$). **g** Net background current (I_b) and its calcium, potassium, and sodium components ($I_{b,Ca}$, $I_{b,K}$ and $I_{b,Na}$, respectively). Data for the Demir et al. and Dokos et al. models were obtained using the cellular open resource (COR) software [27].

model are shown in the leftmost column of Fig. 4. After reinterpretation of the deactivation of the outward 'pacemaker current' I_{K2} of the Bristow–Clark model as an activation of inward I_f [89], the model has four time-dependent (gated) ionic currents (I_{si}, I_K, I_f, and I_{Na}) and a relatively large time-independent (non-gated) 'background current'. The Bristow–Clark model has been updated by Reiner and Antzelevitch [73], who replaced I_{K2} by an I_f current consistent with the data obtained in the first voltage clamp experiments on small SA node preparations. Further, "to generate a biologically accurate action potential", i.e. an action potential with a realistic shape and duration, I_{K1} (shown as $I_{b,K}$ in Fig. 4) was increased by 30%, I_{Na} was decreased by 12% and I_{si} was increased by 15%.

A large body of information on the ionic current systems underlying pacemaker activity had been obtained from voltage clamp experiments on small rabbit SA node preparations containing ≈100 cells carried out in the mid 1970s by Irisawa and Noma (see review by Irisawa et al. [37]). From data obtained in these studies, the first mathematical model of (multicellular) SA node pacemaker activity was constructed by Yanagihara et al. [91]. An extension of this model, based on more recent experimental information on the 'slow inward current' (or 'secondary inward current', I_{si}) was presented by Irisawa and Noma [36]. It has the same five membrane current components as the Bristow–Clark model: I_{si}, I_K, I_f, I_{Na}, and I_b (Fig. 4). However, unlike the Bristow–Clark model, the equations for I_{si}, I_K, and I_f are based on data from voltage clamp experiments on SA node preparations. The I_{Na} equations were adopted from the MNT model, whereas the background current was estimated from the difference between the computed total amplitude of the four gated currents and the experimental steady-state current–voltage curve. Although the action potentials of the Bristow–Clark and Irisawa–Noma models are quite similar (Fig. 4a; Table 3), the underlying ionic currents (Fig. 4c–g) are clearly not. This highlights the different approaches of either modifying model equations to obtain a particular 'reference action potential'—an approach that has recently been followed by Lovell et al. [50]—or incorporating voltage clamp data from SA node preparations with a minimal set of 'adjustable' parameters.

In 1985, DiFrancesco and Noble [14] published a new model of Purkinje fibre electrical activity that, for the first time, fully incorporated the currents generated by the electrogenic sodium–calcium exchanger and sodium–potassium pump (I_{NaCa} and I_{NaK}, respectively) and accounted for the variations in intracellular sodium, calcium, and potassium concentrations. From this model, a model of SA node pacemaker activity was developed by Noble and Noble [61]. The DiFrancesco–Noble equations were used "with parameters appropriate to the SA node except where specific information on the SA node existed that required the equations to be changed". Calcium uptake and release by the SR was represented with separate calcium stores (Fig. 2, top left). The slow inward current was separated into I_{NaCa} and a fast gated component with calcium-dependent inactivation (I_{Ca}). The equations for I_f and I_K were based on voltage clamp experiments on multicellular SA node preparations, whereas the I_{Na} equations were based on data obtained on Purkinje fibres. The equations for I_{NaCa} and for intracellular calcium storage and release were only partly based on experimental observations. The conductance of the calcium background current $I_{b,Ca}$ was chosen to give a diastolic free calcium level in the presumed

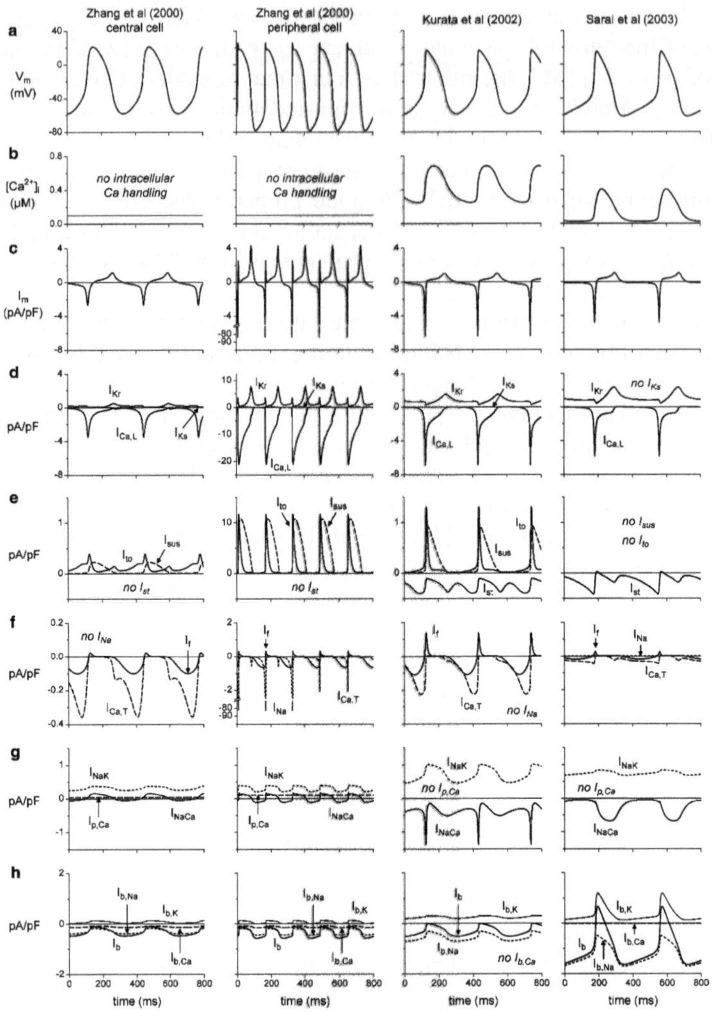

Fig. 6. Action potentials, calcium transients, and transmembrane ionic currents during 800 ms of spontaneous electrical activity in the single rabbit SA nodal cell models of Zhang et al. [93], Kurata et al. [43], and Sarai et al. [75]. **a** Membrane potential (V_m). **b** Intracellular free calcium concentration ($[Ca^{2+}]_i$). **c** Net transmembrane current (I_m). **d** Rapid and slow delayed rectifier potassium current (I_{Kr} and I_{Ks}, respectively) and L-type calcium current ($I_{Ca,L}$). **e** Transient and sustained components of the 4-AP-sensitive outward current (I_{to} and I_{sus}, respectively) and sustained inward current (I_{st}). **f** Fast sodium current (I_{Na}), hyperpolarization-activated current (I_f), and T-type calcium current ($I_{Ca,T}$). **g** Sodium–potassium pump current (I_{NaK}), sodium–calcium exchange current (I_{NaCa}), and sarcolemmal calcium pump current ($I_{p,Ca}$). **h** Net background current (I_b) and its calcium, potassium, and sodium components ($I_{b,Ca}$, $I_{b,K}$, and $I_{b,Na}$, respectively). Note the different ordinate scales and axis breaks for the Zhang et al. peripheral cell model. Models were coded in Visual Fortran 6.6C and carefully checked against the source code that was kindly provided by Dr. Henggui Zhang and Dr. Yasutaka Kurata, or made available on the Kyoto model web site (Table 4).

physiological range of 50–100 nM. The conductance of the potassium background current I_{K1} (shown as $I_{b,K}$ in Fig. 4) was reduced to a small arbitrary value. The conductance of the sodium background current $I_{b,Na}$ was chosen such that a maximum diastolic potential between –70 and –55 mV was achieved. The equation for I_{NaK} was chosen to allow the intracellular sodium and potassium concentrations to be maintained during spontaneous electrical activity. These concentrations show little variation during the course of an action potential and can thus be set to a constant value, avoiding the problems associated with second-generation models (see paragraph 2.4). The 'central' version of the model, mimicking the electrical activity in the centre of the SA node, differs from the standard 'peripheral' version in that I_{Ca} and I_K are reduced by 37.5 and 33%, respectively. The two rightmost columns of Fig. 4 show the action potentials, calcium transients, and associated ionic currents in either version of the model. Because calcium buffering is not incorporated into the model, the intracellular free calcium concentration ($[Ca^{2+}]_i$) can reach levels as high as 7 μM (Fig. 4b). Several ionic currents other than I_{Ca} and I_K are also larger in the peripheral model, e.g. I_f and I_{NaCa}, although the equations are the same in the two models, emphasizing the dependence of these currents on V_m and $[Ca^{2+}]_i$.

Table 3. Action potential parameters of SA nodal cell models

Model	MDP (mV)	APA (mV)	CL (ms)	APD_{50} (ms)	APD_{100} (ms)	dV/dt_{max} (V/s)	DDR (mV/s)
Bristow and Clark [6]	–61	73	361	93	181	2.2	69
Irisawa and Noma [36]	–66	84	329	73	146	5.2	88
Noble and Noble [61] central cell	–61	84	263	70	17	4.7	204
Noble and Noble [61] peripheral cell	–73	102	254	55	122	8.1	247
Noble et al. [62]	–74	106	169	45	80	13.8	439
Wilders et al. [89]	–66	97	388	91	165	7.3	66
Demir et al. [11]	–61	96	263	86	125	9.6	194
Dokos et al. [16]	–65	81	385	78	135	8.8	53
Zhang et al. [93] central cell	–58	79	327	139	214	2.7	183
Zhang et al. [93] peripheral cell	–78	104	161	75	105	83.1	418
Kurata et al. [43]	–59	75	307	107	186	6.4	170
Sarai et al. [75]	–62	77	377	102	171	4.8	97

MDP Maximum diastolic potential, *APA* action potential amplitude, *CL* cycle length, APD_{50} and APD_{100} action potential duration at 50 and 100% repolarization, respectively, dV/dt_{max} maximum upstroke velocity, *DDR* diastolic depolarization rate over 40-ms time interval starting at MDP + 1 mV

3.2 SA Nodal Cell Models of the 1990s: Single Cell Models

The first model of pacemaker activity of a single isolated rabbit SA node pacemaker cell was developed from the multicellular Noble and Noble [61] peripheral model by Noble et al. [13, 62]. The approach was to use the same equations as in the multicellular model with the parameters scaled down appropriately for single cells. Comparing current magnitudes in the multicellular model with experimentally observed single cell current magnitudes suggested that the multicellular model represented ≈100 cells. Therefore, all membrane ionic currents were scaled down by a factor of 100. The single cell was assumed to be a cylinder of 100 μm in length and 8 μm in diameter having a capacitance of 27 pF [13]. New equations for the kinetics of I_f and I_K were formulated from single cell experimental data given by DiFrancesco and Noble [15]. However, I_K had to be scaled up by a factor of 2.7 to evoke pacemaker activity. Despite this the model did not yet provide a quantitatively accurate description of SA node cell activity [13], as illustrated in the leftmost column of Fig. 5 and Table 3.

A more accurate model of single SA node cell activity was developed from the Noble and Noble [61] model by Wilders et al. [89] by incorporating single cell experimental data on cell dimensions, membrane capacitance, I_{Ca} (separated into $I_{Ca,L}$ and $I_{Ca,T}$), I_f, I_K, I_{NaK}, and the amplitude of I_{NaCa}. When sufficient data were not available, which was the case with the intracellular calcium uptake and release processes as well as $I_{b,Ca}$, $I_{b,Na}$, and I_{NaCa}, the model equations were adopted from the Noble and Noble model.

In subsequent years, another two single SA nodal cell models were published. Like the Wilders et al. model [89], the model by Demir et al. [11] is a 'DiFrancesco–Noble-type model'. Except for the inclusion of the sarcolemmal calcium pump current, it has the same set of ionic currents as the Wilders et al. model, partly based on the same voltage clamp data. However, it has more detailed fluid compartment formulations. In particular, there is calcium buffering in the SR through calsequestrin and in the cytoplasm through calmodulin and, to a lesser extent, troponin, which explains the smaller free calcium transient compared to the Noble et al. and Wilders et al. models (Fig. 5b). Furthermore, the extracellular medium is separated into the bulk medium and a restricted cleft space.

The model by Dokos et al. [16] is partly based on the same equations as the Wilders et al. model [89] and Demir et al. [11] model. Unlike the latter models, it does not include the inward calcium background current. Also, it lacks the outward calcium current generated by the sarcolemmal calcium pump. Like the Demir et al. model, the extracellular medium is separated into a restricted space ('extracellular space') and a bulk medium ('vasculature buffer'). There is calcium buffering in the SR, through calsequestrin, but not in the cytoplasm. Although the current traces look similar to those of Dokos et al. [16] (their Fig. 5), the action potential parameters listed in Table 3 are slightly different from those reported by Dokos et al. [16] (their Table 1). In particular, the cycle length of 385 ms is shorter than that reported by Dokos et al. (407 ms). Notably, shorter cycle lengths have also been obtained by Garny et al. [25] using the OxSoft Heart 4.X implementation of the Dokos et al. model (363.6 ms) and by Kurata et al. [43] (385.6 ms).

To simulate the response of SA nodal cells to acetylcholine (ACh), the Demir et al. and Dokos et al. models have both been extended with equations for the muscarinic potassium current ($I_{K,ACh}$) and the modulation of other ionic currents by ACh [12, 17]. In addition, the extended Demir et al. model accounts for modulation of ionic currents by isoprenaline [12].

3.3 SA Nodal Cell Models of the New Millennium: Pacemaker Activity in Greater Detail

After the development of the Wilders et al., Demir et al. and Dokos et al. models, a large body of new experimental data from patch-clamp experiments on SA nodal cells became available. The delayed rectifier current (I_K) was separated into rapidly and slowly activating components (I_{Kr} and I_{Ks}, respectively) [46] and additional ionic currents were identified, including the sustained inward current (I_{st}) [30, 58, 78] and the 4-aminopyridine-sensitive (4-AP-sensitive) outward current with its transient and sustained components (I_{to} and I_{sus}, respectively) [47], albeit with some debate on the latter [85]. It should be noted that I_{st}, I_{to} and I_{sus} are not consistently found in experiments on SA nodal cells.

Furthermore, as reviewed by Boyett et al. [4], a correlation between current density and cell size (membrane capacitance) was reported for several ionic currents in addition to a correlation between action potential shape and cell size, with higher densities and more pronounced action potentials in larger, presumably peripheral cells. Thus, Zhang et al. [93] developed models of 'central' and 'peripheral' SA nodal cells that differ in cell size (20 vs. 65 pF) and current densities, resulting in distinct 'central' and 'peripheral' action potential shapes (two leftmost columns of Fig. 6[1]). The peripheral cell beats at a high rate (6.2 Hz) and its upstroke, with a maximum upstroke velocity of 83 V/s (Table 3), is driven by sodium current rather than L-type calcium current. I_{Ks}, I_{to}, and I_{sus}, but not I_{st}, have been incorporated into the Zhang et al. [93] models. The models have pump and exchanger currents, but no provisions are made for changes in ion concentrations: these are all set to constant values. In particular, $[Ca^{2+}]_i$ is fixed at 100 nM (Fig. 6b). The models have been extended with equations for intracellular calcium handling and sustained inward current by Boyett et al. [5], but thus far these extended models have not been widely used. Another extension of the Zhang et al. models is the incorporation of equations for $I_{K,ACh}$ and the modulation of I_f and $I_{Ca,L}$ by ACh [94].

Building on previous models [11, 16, 89, 93], Kurata et al. [43] formulated an improved model of a primary SA nodal pacemaker cell. The model includes I_{st}, new formulations for voltage and calcium dependent inactivation kinetics of $I_{Ca,L}$, new expressions for activation kinetics of I_{Kr} (see also [44]), revised kinetic formulas for I_{to} and I_{sus}, and new formulations for voltage-and concentration-dependent kinetics of I_{NaK}. Furthermore, the intracellular volume is not only separated into an SR volume, with calcium buffering through calsequestrin, and a myoplasmic volume, with calcium buffering through calmodulin and troponin, but the myoplasm also has a restricted subsarcolemmal space as a diffusion barrier for calcium ions. Unlike some of the earlier models, the Kurata et al. model does not include I_{Na}, $I_{b,Ca}$, and $I_{p,Ca}$: these three currents were assumed to be negligible. As in the models by Zhang et al. [93], I_{Ks} is relatively small (Fig. 6d) and has little effect on normal pacemaker activity, as demonstrated by the <1 ms change in cycle length upon complete block of I_{Ks} (data not shown). The model includes $I_{K,ACh}$, which constitutes the current that appears in Fig. 6h as $I_{b,K}$. The Kurata et al. model was modified by Maltsev et al. [53] to include a phenomenological representation of spontaneous calcium release during diastolic depolarization to assess the functional importance of this calcium release, which could accelerate diastolic depolarization through I_{NaCa}.

Matsuoka et al. [55] developed the 'Kyoto model', which is a compound model that can both simulate a ventricular cell and an SA node cell, and includes a contraction model that allows the calculation of sarcomere shortening. The SA node model, which was further examined in an accompanying paper by Sarai et al. [75],

[1] As set out in detail by Garny et al. [26], the model equations listed in the paper by Zhang et al. [93] do not match the simulation results presented in the same paper. The action potentials and current traces shown in Fig. 6 have been obtained with the correct equations, which are listed by Garny et al. as the '0D capable' version of the models [26].

and is therefore referred to as the Sarai et al. model, can be obtained from the ventricular cell model by adjusting current densities and membrane capacitance as well as the cell volume and the relative volumes of the calcium uptake and release stores. Furthermore, the concentration of troponin in the SA nodal cell is 10% of that in the ventricular cell. The Kyoto model does not include equations for $I_{p,Ca}$ and I_{sus}. The model does have equations for I_{Ks} and I_{to}, but these two currents are set to zero in the SA node variant of the Kyoto model. The model has five background currents: (1) a non-selective cation current with K^+ and Na^+ components, (2) a plateau potassium current, (3) a calcium-activated non-selective cation current with K^+ and Na^+ components, (4) an ATP-sensitive potassium current, and (5) a small calcium background current. Together with the small inward rectifier current (I_{K1}) and $I_{K,ACh}$ that are also included in the Kyoto model, these currents constitute the current that appears in Fig. 6h as I_b.

The action potentials of the Kurata et al. and Sarai et al. models are quite similar (Fig. 6a; Table 3). However, the diastolic interval and cycle length of the two models are remarkably different. This also holds for the calcium transients of the two models, with $[Ca^{2+}]_i$ fluctuations between 247 and 683 nM in the Kurata et al. model and 23 and 407 nM in the Sarai et al. model (Fig. 6b). Furthermore, there are clear differences in the magnitude and time course of some of the ionic currents (Fig. 6c–h).

3.4 Caveats

The diastolic depolarization phase of the SA nodal action potential is driven by a net ionic current (I_m in Figs. 4, 5, 6) that is the balanced result of a large number of inward and outward currents. The small diastolic I_m can result from many different combinations of the same set of currents that interact in a complex way. For example, as shown by both Noble et al. [64] and Wilders et al. [89], $I_{b,Na}$ and I_f act in a reciprocal way, and the effect of block of I_f depends on the amount of $I_{b,Na}$ incorporated into the model. This demonstrates why the amount of slowing of pacemaker activity upon block of a particular inward current, which is often used to quantify the modulatory capacity of a current, is poorly predictable. This is illustrated in Fig. 7, which shows the percent increase in cycle length upon complete block of either the native $I_{Ca,T}$ (hatched bars), if present, or the native I_f (grey bars) of each of the 12 SA nodal cell models presented in Figs. 4, 5, and 6. The increase in cycle length upon block of $I_{Ca,T}$ ranges from 2.4% (Wilders et al. model and Sarai et al. model) to 24% (Zhang et al. central cell), whereas the increase upon block of I_f ranges from 0.9% (Sarai et al. model) to 30% (Zhang et al. peripheral cell). A model that is sensitive to the block of $I_{Ca,T}$ is not necessarily sensitive to the block of I_f and vice versa. Figure 7 clearly demonstrates that one should be very careful in drawing firm conclusions from computer simulations, e.g. regarding the role of $I_{Ca,T}$ in pacemaker activity.

The 20.4% increase in cycle length upon block of $I_{Ca,T}$ and 4.8% upon block of I_f in the Demir et al. model shown in Fig. 7—similar values have been reported for the Demir et al. model by Kurata et al. [43]—are considerably smaller than the values of 26 and 20% reported in the original paper by Demir et al. [11] (their Figs. 14, 15). The data shown in Fig. 7 have been obtained under quasi-steady-state conditions, i.e. after running the model for a sufficiently long time to reach quasi-steady-state behaviour, whereas the data by Demir et al. [11] are most likely taken from a short train of action potentials that have yet not reached state, as can be readily checked by running the 'iCell' version of the model (see paragraph 4.1). This underscores the importance of running models for a sufficiently long time to reach quasi-steady-state behaviour to avoid overestimating (or underestimating) the role of particular ionic currents in pacemaker activity.

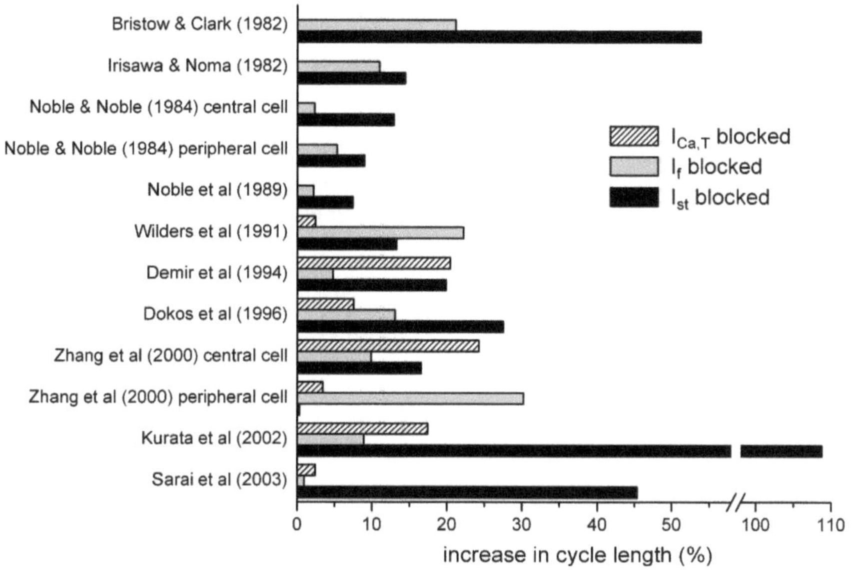

Fig. 7. Percent increase in cycle length upon complete block of T-type calcium current ($I_{Ca,T}$), hyperpolarization-activated current (I_f), or sustained inward current (I_{st}) in various rabbit SA node models. In case of $I_{Ca,T}$ and I_f, the conductance of the native current, if present, was set to zero. In case of I_{st}, this current was introduced into each of the models using the equations of Shinagawa et al. [78], with the I_{st} conductance set to 35.6 pA/pF [95]. In the models by Kurata et al. and Sarai et al., this I_{st} replaced the native I_{st}, which changed the intrinsic cycle length of these models from 307 to 298 ms and from 377 to 342 ms, respectively.

The differences in response to block of $I_{Ca,T}$ or I_f among the 12 SA nodal cell models may reflect differences in the current density or kinetics of these currents between these models. Therefore, it was also tested how each of the models responds to block of a current that is incorporated into each of the models with the same density and kinetics. To this end, the Shinagawa et al. [78] I_{st} equations were incorporated into each of the models. In the two models with a native I_{st}, i.e. the Kurata et al. and Sarai et al. models, the native I_{st} equations were replaced with those of Shinagawa et al. [78], which are also listed in the minireview of Ist by Mitsuiye et al. [58]. The maximum I_{st} conductance was set to 35.6 pS/pF, as used by Zhang et al. in a study in which they incorporated the Shinagawa et al. I_{st} [78] in six different SA nodal cell models to investigate how the beating rate of each model increased upon the incorporation of I_{st} [95]. Zhang et al. [95] found that incorporation of I_{st} increased the beating rate by 2.2% in the Noble–Noble central cell model, by 14% in the

Wilders et al. model, by 20% in the Demir et al. model, by 0.8% in the Dokos et al. model, and by 7 and 0.6% in the Zhang et al. central and peripheral cell models, respectively. They concluded that the importance of I_{st}, in terms of its effect on beating rate, depends on its amplitude relative to the amplitude of other inward currents that are active during diastolic depolarization [95]. This is corroborated by the results shown in Fig. 7.[2] The percent increase in cycle length upon block of I_{st}, which is equivalent to the percent increase in beating rate upon incorporation of I_{st}, is generally smaller in the models that have relatively large diastolic inward currents like $I_{b,Na}$, $I_{Ca,T}$, I_f, or I_{Na} (cf. Figs. 4, 5, 6). The increase in cycle length upon block of I_{st} ranges from 0.3% for the Zhang et al. peripheral cell model to 108% for the Kurata et al. model (Fig. 7), although the same set of equations was incorporated into each of the 12 models. This emphasizes the limitations of investigating the role of a novel ionic current in pacemaker activity through the use of a single SA nodal cell model. In the particular case of I_{st}, for which a pharmacological blocker is not available, the dynamic clamp method could be used to current-clamp an isolated SA nodal cell and subtract I_{st} in real time, thus establishing an 'electronic block' of the current [87].

4 Tools for Running Cardiac Cell Models

4.1 Public Tools

The source code of several cardiac cell models, including some SA nodal cell models, is publicly available through the Internet. This source code can be compiled into a computer program using the appropriate 'compiler', which requires some computer programming skills. Generally, the resulting program is a fast-running console application with minimal input and output rather than a flexible program with a graphical user interface. Expertise in computer modelling is required to tailor the program to personal needs. The public availability of source code can help identify (typographical) errors in model equations and prevent erroneous implementations of published models. Table 4 lists Internet sites at which source codes are made available. Apart from 'CellML' code (see below), the availability of source code of SA nodal cell models is limited to the central and peripheral models by Zhang et al. [93, 94] and the Sarai et al. model [75]. The Zhang et al. models are available as Fortran code, whereas the Kyoto model, which includes the Sarai et al. model, is available in various formats, including C++, Visual Basic, Delphi 7, and MATLAB.

Several efforts have been made to set a language standard for cardiac cell modelling. One example is the Mathematical Modeling Language (MML) for use with 'JSim' (for Java Simulator), which is a freely available Java-based simulation system for building and analysing quantitative numeric models with its primary focus in physiology and

[2] There is a discrepancy between the percentages shown in Fig. 7 and those reported by Zhang et al. [95]. The most striking one is in the percentage obtained with the Dokos et al. model: 27.5% in Fig. 7 versus 0.8% in the study by Zhang et al. [95]. The latter discrepancy seems to be related to an error in current amplitudes in the study by Zhang et al. [95], e.g. in the sodium background current, which has a peak amplitude >100 nA in their implementation of the Dokos et al. model (their Fig. 3B), whereas the correct value is <40 pA (Fig. 5 of Ref. [16]).

biomedicine. JSim downloads are available for the Windows, Macintosh, and Linux platforms from the National Simulation Resource at the University of Washington (http://nsr.bioeng.washington.edu/). Another example is the 'CellML' language,

Table 4. Cardiac cell models available on the Internet

Source code
Source code of numerous models in CellML format [31, 49]
Web address: http://www.cellml.org/models/
Source code of cardiac cell models from the Rudy group (C++, MATLAB)
Web address: http://www.rudylab.wustl.edu/
Source code of cardiac cell models from the Winslow group (C++, Fortran)
Web address: http://www.ccbm.jhu.edu/
Source code of the 'Kyoto model' [55, 75] (several formats)
Web address: http://www.card.med.kyoto-u.ac.jp/simulation/
C++ source code of the Bernus et al. human ventricular cell model [2]
Web address: http://www.physiol.med.uu.nl/wilders/
C++ source code of the ten Tusscher et al. human ventricular cell models [80, 81]
Web address: http://www-binf.bio.uu.nl/khwjtuss/
Fortran source code of the Zhang et al. rabbit sinoatrial cell models [93, 94]
Web address: http://www.personalpages.manchester.ac.uk/staff/H.Zhang-3/
Ready-to-use Java applets
Java applets of numerous ionic models [22]
Web address: http://www.arrhythmia.hofstra.edu/
iCell: Java-based interactive cell modelling resource [10]
Web address: http://www.ssd1.bme.memphis.edu/icell/
simBio: Java package for 'biological dynamic simulation' [76]
Web address: http://www.sim-bio.org/
Ready-to-use Windows programs
COR (cellular open resource): comes with various ionic models [27]
Web address: http://www.cor.physiol.ox.ac.uk/
LabHEART: Windows version of Puglisi–Bers rabbit ventricular cell model [71]
Web address: http://www.meddean.luc.edu/templates/ssom/depts/physio/labheart.cfm

See Table 1 of Ref. [76] for a comprehensive list of tools for cardiac cell modelling

which is an open standard based on the XML markup language, being developed by the Bioengineering Institute at the University of Auckland and affiliated research groups [31, 49]. The purpose of CellML is to provide a means for unambiguous specification of biological models, which can be used to store and exchange such models. The structure of a CellML document is such that it can both be used to generate equations suitable for publishing and to generate a computer code to perform simulations, thus ensuring that published equations and simulation results are consistent. To run simulations, a simulation environment that imports CellML files and compiles these into executable computer code is required. Two such simulation environments are the afore-mentioned JSim simulation system and 'COR' (cellular open source), which runs under the Windows operating system [27].

A large number of cardiac cell models are already available in the 'model repository' on the CellML web site (Table 4). This includes the SA nodal cell models by Noble and Noble [61], Demir et al. [11, 12], Dokos et al. [16, 17], Zhang et al. [93], Boyett et al. [5], Kurata et al. [43], Sarai et al. [75], and Lovell et al. [50]. Unfortunately, once imported into JSim or COR, the far majority of these CellML files generate various errors at compilation time, whereas the few CellML files that can be compiled into executable computer code either generate errors at runtime or result in erroneous model output. Similar problems are encountered when using a third simulation environment formed by the Firefox web browser with the mozCellML plugin, which is provided on the CellML web site. Many of these errors seem to result from equations with unbalanced units, i.e. equations in which the left-hand and right-hand sides have different units. Currently, efforts are starting to be made to ensure that the models that are posted on the CellML web site have been tested for consistency and completeness.

Fortunately, the COR simulation environment comes with a collection of CellML files that all compile and run properly. This collection includes the SA nodal cell models by Noble and Noble [61], Demir et al. [11], Dokos et al. [16], and Zhang et al. [93], and an updated version of the model by Noble et al. [62]. With the included CellML files, the COR simulation environment forms very flexible Windows based software with a nice graphical user interface that can be readily used by non-experts in computer modelling to run cardiac cell models. Model parameters can easily be changed through the user interface and model equations can be changed in the CellML editor.

As listed in Table 4, more cardiac cell models are available for use by non-experts in computer modelling. The rabbit ventricular cell model by Puglisi and Bers [71] is available as a stand-alone Windows program. Furthermore, various cardiac cell models, including several SA nodal cell models, are available as Java applets, i.e. programs that can be run from within a web browser. These applets come with limitations. In some, steady-state behaviour cannot be reached within the maximum duration of the simulation. In others, it is not possible to save data or change specific parameters. Nevertheless, the applets provide the user with a rapid 'look and feel' of various cardiac cell models.

4.2 Commercial Solutions

In addition to the publicly available source code and software listed in Table 4, there are a few commercially available tools for running cardiac cell models. The well-known OxSoft HEART software, which was released in 1984 and has evolved from a pure text based DOS application into an application with a proper graphical user interface, is no longer being developed due to technical reasons that prevent programs written in Borland Pascal from running on the latest generation of processors [27]. The COR software by Dr. Alan Garny (see paragraph 4.1), which contains several of the models that are available in the OxSoft HEART package, is now freely available to the scientific community. Recently, Simulogic Inc (Halifax, Canada; http://www.simulogic.com/) started to sell published cardiac cell models in a format that is ready for use with the open-source Java-based cell electrophysiology simulation environment (CESE). The 'academic pricing' as of April 2007 is at US $ 350 for a single cardiac cell model.

5 Concluding Remarks

Today's cardiac cell models have a common structure that was introduced with the Nobel Prize awarded studies by Hodgkin and Huxley. Over the course of 25 years, SA nodal cell models have evolved from 'simple' models with four time-dependent sarcolemmal ionic currents and seven state variables, i.e. the membrane potential and six gating variables, to comprehensive models with >40 state variables that not only keep track of transmembrane potential and sarcolemmal ionic currents, but also account for ion concentrations in intracellular and extracellular compartments. Hopefully, this review has clarified some of the merits and limitations of current SA nodal cell models and the tools that are available for running these models.

References

1. Beeler GW, Reuter H (1977) Reconstruction of the action potential of ventricular myocardial fibres. J Physiol 268:177– 210
2. Bernus O, Wilders R, Zemlin CW, Verschelde H, Panfilov AV (2002) A computationally efficient electrophysiological model of human ventricular cells. Am J Physiol Heart Circ Physiol 282:H2296–H2308. DOI 10.1152/ajpheart.00731.2001
3. Bondarenko VE, Szigeti GP, Bett GC, Kim SJ, Rasmusson RL (2004) Computer model of action potential of mouse ventricular myocytes. Am J Physiol Heart Circ Physiol 287:H1378–H1403. DOI 10.1152/ajpheart.00185.2003
4. Boyett MR, Honjo H, Kodama I (2000) The sinoatrial node, a heterogeneous pacemaker structure. Cardiovasc Res 47:658–687. DOI 10.1016/S0008-6363(00)00135-8
5. Boyett MR, Zhang H, Garny A, Holden AV (2001) Control of the pacemaker activity of the sinoatrial node by intracellular Ca^{2+}. Experiments and modeling. Philos Trans R Soc Lond A Math Phys Eng Sci 359:1091–1110. DOI 10.1098/ rsta.2001.0818
6. Bristow DG, Clark JW (1982) A mathematical model of primary pacemaking cell in SA node of the heart. Am J Physiol 243:H207–H218
7. Cabo C, Boyden PA (2003) Electrical remodeling of the epicardial border zone in the canine infarcted heart: a computational analysis. Am J Physiol Heart Circ Physiol 284:H372–H384. DOI 10.1152/ajpheart.00512.2002
8. Courtemanche M, Ramirez RJ, Nattel S (1998) Ionic mechanisms underlying human atrial action potential properties: insights from a mathematical model. Am J Physiol 275:H301–H321
9. de Boer TP, van Veen TAB, Houtman MJC, Jansen JA, van Amersfoorth SCM, Doevendans PA, Vos MA, van der Heyden MAG (2006) Inhibition of cardiomyocyte automaticity by electrotonic application of inward rectifier current from Kir2.1 expressing cells. Med Biol Eng Comput 44:537–542. DOI 10.1007/s11517-006-0059-8
10. Demir SS (2006) Interactive cell modeling webresource, iCell, as a simulation-based teaching and learning tool to supplement electrophysiology education. Ann Biomed Eng 34:1077–1087. DOI 10.1007/s10439-006-9138-0

11. Demir SS, Clark JW, Murphey CR, Giles WR (1994) A mathematical model of a rabbit sinoatrial node cell. Am J Physiol 266:C832–C852
12. Demir SS, Clark JW, Giles WR (1999) Parasympathetic modulation of sinoatrial node pacemaker activity in rabbit heart: a unifying model. Am J Physiol 276:H2221–H2244
13. Denyer JC (1989) Isolation and electrophysiological characteristics of rabbit sino-atrial node cells. Ph.D. thesis, University of Oxford
14. DiFrancesco D, Noble D (1985) A model of cardiac electrical activity incorporating ionic pumps and concentration changes. Philos Trans R Soc Lond B Biol Sci 307:353–398
15. DiFrancesco D, Noble D (1989) Current I_f and its contribution to cardiac pacemaking. In: Jacklet JW (ed) Neuronal and cellular oscillators. Marcel Dekker, New York, pp 31–57. ISBN 0-8247-8030-2
16. Dokos S, Celler B, Lovell N (1996) Ion currents underlying sinoatrial node pacemaker activity: a new single cell mathematical model. J Theor Biol 181:245–272. DOI 10.1006/jtbi.1996.0129
17. Dokos S, Celler BG, Lovell NH (1996) Vagal control of sinoatrial rhythm: a mathematical model. J Theor Biol 182:21–44. DOI 10.1006/jtbi.1996.0141
18. Drouhard JP, Roberge FA (1987) Revised formulation of the Hodgkin–Huxley representation of the sodium current in cardiac cells. Comput Biomed Res 20:333–350. DOI 10.1016/ 0010-4809(87)90048-6
19. Earm YE, Noble D (1990) A model of the single atrial cell: relation between calcium current and calcium release. Proc R Soc Lond B Biol Sci 240:83–96
20. Endresen LP, Hall K, Høye JS, Myrheim J (2000) A theory for the membrane potential of living cells. Eur Biophys J 29:90–103. DOI 10.1007/s002490050254
21. Fenton F, Karma A (1998) Vortex dynamics in three-dimensional continuous myocardium with fiber rotation: filament instability and fibrillation. Chaos 8:20–47. DOI 10.1063/1.166311
22. Fenton FH, Cherry EM, Hastings HM, Evans SJ (2002) Real-time computer simulations of excitable media: JAVA as a scientific language and as a wrapper for C and FORTRAN programs. Biosystems 64:73–96. DOI 10.1016/S0303-2647(01)00177-0
23. FitzHugh R (1961) Impulses and physiological states in theoretical models of nerve membrane. Biophys J 1:445–466
24. Fox JJ, McHarg JL, Gilmour RF Jr (2002) Ionic mechanism of electrical alternans. Am J Physiol Heart Circ Physiol 282:H516–H530. DOI 10.1152/ajpheart.00612.2001
25. Garny A, Noble PJ, Kohl P, Noble D (2002) Comparative study of rabbit sino-atrial node cell models. Chaos Solitons Fractals 13:1623–1630. DOI 10.1016/S0960-0779(01) 00171-0
26. Garny A, Kohl P, Hunter PJ, Boyett MR, Noble D (2003) One-dimensional rabbit sinoatrial node models: benefits and limitations. J Cardiovasc Electrophysiol 14:S121–S132. DOI 10.1046/j.1540.8167.90301.x
27. Garny A, Kohl P, Noble D (2003) Cellular open resource (COR): a public CellML based environment for modelling biological function. Int J Bifurcat Chaos 13:3579–3590. DOI 10.1142/S021812740300882X
28. Greenstein JL, Winslow RL (2002) An integrative model of the cardiac ventricular myocyte incorporating local control of Ca^{2+} release. Biophys J 83:2918–2945
29. Guevara MR, Lewis TJ (1995) A minimal single-channel model for the regularity of beating in the sinoatrial node. Chaos 5:174–183. DOI 10.1063/1.166065
30. Guo J, Ono K, Noma A (1995) A sustained inward current activated at the diastolic potential range in rabbit sino-atrial node cells. J Physiol 483:1–13

31. Hedley WJ, Nelson MR, Bullivant DP, Nielsen PF (2001) A short introduction to CellML. Philos Trans R Soc Lond A Math Phys Eng Sci 359:1073–1089. DOI 10.1098/rsta.2001.0817
32. Henriquez CS (1993) Simulating the electrical behavior of cardiac tissue using the bidomain model. Crit Rev Biomed Eng 21:1–77
33. Hilgemann DW, Noble D (1987) Excitation–contraction coupling and extracellular calcium transients in rabbit atrium: reconstruction of the basic cellular mechanisms. Proc R Soc Lond B Biol Sci 230:163–205
34. Hodgkin AL, Huxley AF (1952) A quantitative description of membrane current and its application to conduction and excitation in nerve. J Physiol 117:500–544
35. Hund TJ, Rudy Y (2004) Rate dependence and regulation of action potential and calcium transient in a canine cardiac ventricular cell model. Circulation 110:3168–3174. DOI 10.1161/01.CIR.0000147231.69595.D3
36. Irisawa H, Noma A (1982) Pacemaker mechanisms of rabbit sinoatrial node cells. In: Bouman LN, Jongsma HJ (eds) Cardiac rate and rhythm: physiological, morphological, and developmental aspects. Martinus Nijhoff, London, pp 35–51. ISBN 90-247-2626-3
37. Irisawa H, Brown HF, Giles W (1993) Cardiac pacemaking in the sinoatrial node. Physiol Rev 73:197–227
38. Iyer V, Mazhari R, Winslow RL (2004) A computational model of the human left-ventricular epicardial myocyte. Biophys J 87:1507–1525. DOI 10.1529/ biophysj.104.043299
39. Jafri S, Rice JR, Winslow RL (1998) Cardiac Ca^{2+} dynamics: the roles of ryanodine receptor adaptation and sarcoplasmic reticulum load. Biophys J 74:1149–1168
40. Joyner RW, Wilders R, Wagner MB (2006) Propagation of pacemaker activity. Med Biol Eng Comput. 45:177–187. DOI 10.1007/ s11517-006-0102-9
41. Kneller J, Ramirez RJ, Chartier D, Courtemanche M, Nattel S (2002) Time-dependent transients in an ionically based mathematical model of the canine atrial action potential. Am J Physiol Heart Circ Physiol 282:H1437–H1451. DOI 10.1152/ajpheart.00489.2001
42. Krogh-Madsen T, Schaffer P, Skriver AD, Taylor LK, Pelzmann B, Koidl B, Guevara MR (2005) An ionic model for rhythmic activity in small clusters of embryonic chick ventricular cells. Am J Physiol Heart Circ Physiol 289:H398–H413. DOI 10.1152/ajpheart.00683.2004
43. Kurata Y, Hisatome I, Imanishi S, Shibamoto T (2002) Dynamical description of sinoatrial node pacemaking: improved mathematical model for primary pacemaker cell. Am J Physiol Heart Circ Physiol 283:H2074–H2101. DOI 10.1152/ajpheart.00900.2001
44. Kurata Y, Hisatome I, Imanishi S, Shibamoto T (2003) Roles of L-type Ca^{2+} and delayed rectifier K^+ currents in sinoatrial node pacemaking: insights from stability and bifurcation analyses of a mathematical model. Am J Physiol Heart Circ Physiol 285:H2804–H2819. DOI 10.1152/ajpheart.01050.2002
45. Kurata Y, Hisatome I, Matsuda H, Shibamoto T (2005) Dynamical mechanisms of pacemaker generation in I_{K1}-downregulated human ventricular myocytes: insights from bifurcation analyses of a mathematical model. Biophys J 89:2865–2887. DOI 10.1529/biophysj.105.060830
46. Lei M, Brown HF (1996) Two components of the delayed rectifier potassium current, I_K, in rabbit sino-atrial node cells. Exp Physiol 81:725–741
47. Lei M, Honjo H, Kodama I, Boyett MR (2000) Characterisation of the transient outward K^+ current in rabbit sinoatrial node cells. Cardiovasc Res 46:433–441. DOI 10.1016/S00086363(00)00036-5

48. Lindblad DS, Murphey CR, Clark JW, Giles WR (1996) A model of the action potential and underlying membrane currents in a rabbit atrial cell. Am J Physiol 271:H1666– H1691
49. Lloyd CM, Halstead MD, Nielsen PF (2004) CellML: its future, present and past. Prog Biophys Mol Biol 85:433–450. DOI 10.1016/j.pbiomolbio.2004.01.004
50. Lovell NH, Cloherty SL, Celler BG, Dokos S (2004) A gradient model of cardiac pacemaker myocytes. Prog Biophys Mol Biol 85:301–323. DOI 10.1016/j.pbiomolbio.2003.12.001
51. Luo C-H, Rudy Y (1991) A model of the ventricular cardiac action potential: depolarization, repolarization and their interaction. Circ Res 68:1501–1526
52. Luo C-H, Rudy Y (1994) A dynamic model of the cardiac ventricular action potential, I: simulations of ionic currents and concentration changes. Circ Res 74:1071–1096
53. Maltsev VA, Vinogradova TM, Bogdanov KY, Lakatta EG, Stern MD (2004) Diastolic calcium release controls the beating rate of rabbit sinoatrial node cells: numerical modeling of the coupling process. Biophys J 86:2596–2605
54. Mangoni ME, Traboulsie A, Leoni AL, Couette B, Marger L, Le Quang K, Kupfer E, Cohen-Solal A, Vilar J, Shin HS, Escande D, Charpentier F, Nargeot J, Lory P (2006) Bradycardia and slowing of the atrioventricular conduction in mice lacking $Ca_V3.1/\alpha_{1G}$ T-type calcium channels. Circ Res 98:1422–1430. DOI 10.1161/01.RES.0000225862.14314.49
55. Matsuoka S, Sarai N, Kuratomi S, Ono K, Noma A (2003) Role of individual ionic current systems in ventricular cells hypothesized by a model study. Jpn J Physiol 53:105–123. DOI 10.2170/jjphysiol.53.105
56. Mazhari R, Greenstein JL, Winslow RL, Marban E, Nuss HB (2001) Molecular interactions between two long-QT syndrome gene products, HERG and KCNE2, rationalized by in vitro and in silico analysis. Circ Res 89:33–38. DOI 10.1161/hh1301.093633
57. McAllister RE, Noble D, Tsien RW (1975) Reconstruction of the electrical activity of cardiac Purkinje fibres. J Physiol 251:1–59
58. Mitsuiye T, Shinagawa Y, Noma A (2000) Sustained inward current during pacemaker depolarization in mammalian sinoatrial node cells. Circ Res 87:88–91
59. Noble D (1960) Cardiac action and pacemaker potentials based on the Hodgkin–Huxley equations. Nature 188:495–497
60. Noble D (1962) A modification of the Hodgkin–Huxley equations applicable to Purkinje fibre action and pacemaker potentials. J Physiol 160:317–352
61. Noble D, Noble SJ (1984) A model of sino-atrial node electrical activity based on a modification of the DiFrancesco–Noble (1984) equations. Proc R Soc Lond B Biol Sci 222:295–304
62. Noble D, DiFrancesco D, Denyer JC (1989) Ionic mechanisms in normal and abnormal cardiac pacemaker activity. In: Jacklet JW (ed) Neuronal and cellular oscillators. Marcel Dekker, New York, pp 59–85. ISBN 0-8247-8030-2
63. Noble D, Noble SJ, Bett GC, Earm YE, Ho WK, So IK (1991) The role of sodium–calcium exchange during the cardiac action potential. Ann NY Acad Sci 639:334–353
64. Noble D, Denyer JC, Brown HF, DiFrancesco D (1992) Reciprocal role of the inward currents $I_{b,Na}$ and I_f in controlling and stabilizing pacemaker frequency of rabbit sino-atrial node cells. Proc R Soc Lond B Biol Sci 250:199–207
65. Noble D, Varghese A, Kohl P, Noble P (1998) Improved guineapig ventricular cell model incorporating a diadic space, i_{Kr} and i_{Ks}, and length- and tension-dependent processes. Can J Cardiol 14:123–134

66. Nordin C (1993) Computer model of membrane current and intracellular Ca^{2+} flux in the isolated guinea pig ventricular myocyte. Am J Physiol 265:H2117–H2136
67. Nygren A, Fiset C, Firek L, Clark JW, Lindblad DS, Clark RB, Giles WR (1998) Mathematical model of an adult human atrial cell: the role of K^+ currents in repolarization. Circ Res 82:63–81
68. Pandit SV, Clark RB, Giles WR, Demir SS (2001) A mathematical model of action potential heterogeneity in adult rat left ventricular myocytes. Biophys J 81:3029–3051
69. Potse M, Dubé B, Richer J, Vinet A, Gulrajani RM (2006) A comparison of monodomain and bidomain reaction-diffusion models for action potential propagation in the human heart. IEEE Trans Biomed Eng 53:2425–2435. DOI 10.1109/ TBME.2006.880875
70. Priebe L, Beuckelmann DJ (1998) Simulation study of cellular electric properties in heart failure. Circ Res 82:1206–1223
71. Puglisi JL,Bers DM (2001) LabHEART: an interactive computer model of rabbit ventricular myocyte ion channels and Ca transport. Am J Physiol Cell Physiol 281:C2049–C2060
72. Ramirez RJ, Nattel S, Courtemanche M (2000) Mathematical analysis of canine atrial action potentials: rate, regional factors, and electrical remodeling. Am J Physiol Heart Circ Physiol 279:H1767–H1785
73. Reiner VS, Antzelevitch C (1985) Phase resetting and annihilation in a mathematical model of sinus node. Am J Physiol 249:H1143–H1153
74. Rosen MR, Brink PR, Cohen IS, Robinson RB (2006) Biological pacemakers based on I_f. Med Biol Eng Comput 45:157–166. DOI 10.1007/s11517-006-0060-2
75. Sarai N, Matsuoka S, Kuratomi S, Ono K, Noma A (2003) Role of individual ionic current systems in the SA node hypothesized by a model study. Jpn J Physiol 53:125–134. DOI 10.2170/jjphysiol.53.125
76. Sarai N, Matsuoka S, Noma A (2006) SimBio: a Java package for the development of detailed cell models. Prog Biophys Mol Biol 90:360–377. DOI 10.1016/ j.pbiomolbio.2005.05.008
77. Shannon TR, Wang F, Puglisi J, Weber C, Bers DM (2004) A mathematical treatment of integrated Ca dynamics within the ventricular myocyte. Biophys J 87:3351–3371. DOI 10.1529/biophysj.104.047449
78. Shinagawa Y, Satoh H, Noma A (2000) The sustained inward current and inward rectifier K^+ current in pacemaker cells dissociated from rat sinoatrial node. J Physiol 523:593–605
79. Silva J, Rudy Y (2003) Mechanism of pacemaking in I_{K1}-downregulated myocytes. Circ Res 92:261–263. DOI 10.1161/ 01.RES.0000057996.20414.C6
80. ten Tusscher KHWJ, Panfilov AV (2006) Alternans and spiral breakup in a human ventricular tissue model. Am J Physiol Heart Circ Physiol 291:H1088–H1100. DOI 10.1152/ ajpheart.00109.2006
81. ten Tusscher KHWJ, Noble D, Noble PJ, Panfilov AV (2004) A model for human ventricular tissue. Am J Physiol Heart Circ Physiol 286:H1573–H1589. DOI 10.1152/ ajpheart.00794.2003
82. ten Tusscher KHWJ, Bernus O, Hren R, Panfilov AV (2006) Comparison of electro physiological models for human ventricular cells and tissues. Prog Biophys Mol Biol 90:326–345. DOI 10.1016/j.pbiomolbio.2005.05.015
83. van Capelle FJL, Durrer D (1980) Computer simulation of arrhythmias in a network of coupled excitable elements. Circ Res 47:454–466
84. van der Pol B, van der Mark J (1928) The heartbeat considered as a relaxation oscillation and an electrical model of the heart. Philos Mag 6:763–775

85. Verkerk AO, van Ginneken ACG (2001) Considerations in studying the transient outward K^+ current in cells exhibiting the hyperpolarization-activated current. Cardiovasc Res 52:517–520. DOI 10.1016/S0008-6363(01)00482-5
86. Viswanathan PC, Coles JA Jr., Sharma V, Sigg DC (2006) Recreating an artificial biological pacemaker: insights from a theoretical model. Heart Rhythm 3:824–831. DOI 10.1016/j.hrthm.2006.03.012
87. Wilders R (2006) Dynamic clamp: a powerful tool in cardiac electrophysiology. J Physiol 576:349–359. DOI 10.1113/jphysiol.2006.115840
88. Wilders R, Jongsma HJ (1993) Beating irregularity of single pacemaker cells isolated from the rabbit sinoatrial node. Biophys J 65:2601–2613
89. Wilders R, Jongsma HJ, van Ginneken ACG (1991) Pacemaker activity of the rabbit sinoatrial node: a comparison of mathematical models. Biophys J 60:1202–1216
90. Winslow RL, Rice J, Jafri S, Marbán E, O'Rourke B (1999) Mechanisms of altered excitation-contraction coupling in canine tachycardia-induced heart failure, II: model studies. Circ Res 84:571–586
91. Yanagihara K, Noma A, Irisawa H (1980) Reconstruction of sino-atrial node pacemaker potential based on the voltage clamp experiments. Jpn J Physiol 30:841–857
92. Zaniboni M, Pollard AE, Yang L, Spitzer KW (2000) Beat-to-beat repolarization variability in ventricular myocytes and its suppression by electrical coupling. Am J Physiol Heart Circ Physiol 278:H677–H687
93. Zhang H, Holden AV, Kodama I, Honjo H, Lei M, Varghese T, Boyett MR (2000) Mathematical models of action potentials in the periphery and center of the rabbit sinoatrial node. Am J Physiol Heart Circ Physiol 279:H397–H421
94. Zhang H, Holden AV, Noble D, Boyett MR (2002) Analysis of the chronotropic effect of acetylcholine on sinoatrial node cells. J Cardiovasc Electrophysiol 13:465–474. DOI 10.1046/j.1540-8167.2002.00465.x
95. Zhang H, Holden AV, Boyett MR (2002) Sustained inward current and pacemaker activity of mammalian sinoatrial node. J Cardiovasc Electrophysiol 13:809–812. DOI 10.1046/j.1540-8167.2002.00809.x

Application of Mesenchymal Stem Cell-Derived Cardiomyocytes as Bio-pacemakers: Current Status and Problems to Be Solved

Yuichi Tomita[1,2], Shinji Makino[1], Daihiko Hakuno[1,2], Naoichiro Hattan[1], Kensuke Kimura[1], Shunichiro Miyoshi[2], Mitsushige Murata[1,2], Masaki Ieda[1,2], and Keiichi Fukuda[1]

[1] Department of Regenerative Medicine and Advanced Cardiac Therapeutics, Keio University School of Medicine, Institute of Integrated Medical Research, 35 Shinanomachi, Shinjuku-ku, Tokyo 160-8582, Japan
kfukuda@sc.itc.keio.ac.jp

[2] Cardiology Division, Department of Internal Medicine, Keio University School of Medicine, 35 Shinanomachi, Shinjuku, Tokyo 160-8582, Japan

Abstract. Bone marrow mesenchymal stem cells (CMG cells) are multipotent and can be induced by 5-azacytidine to differentiate into cardiomyocytes. We characterized the electrophysiological properties of these cardiomyocytes and investigated their potential for use as transplantable bio-pacemakers. After differentiation, action potentials in spontaneously beating cardiomyocytes were initially sinus node-like, but subsequently became ventricular cardiomyocyte-like. RT-PCR established that ion channels mediating I_{K1} and I_{Kr} were expressed before differentiation. After differentiation, ion channels underlying $I_{Ca,L}$ and I_f were expressed first, followed by ion channels mediating I_{to} and $I_{K,ATP}$. Differentiated CMG cells expressed β-adrenergic receptors and increased their beat rate in response to isoproterenol. CMG cardiomyocytes were purified using GFP fluorescence and transplanted into the free walls of the left ventricles of mice. The transplanted cardiomyocytes survived and connected to surrounding recipient cardiomyocytes via intercalated discs. Although further innovation is required, the present findings provide evidence of the potential for bone marrow-derived cardiomyocytes to be used as bio-pacemakers.

Keywords: Pacemaker, Bone marrow, Mesenchymal, stem cell, Cardiomyocyte Transplantation.

1 Introduction

The use of artificial pacemakers for treatment of bradyarrhythmias such as sick sinus syndrome and atrioventricular block is well established. Although artificial pacemakers are considered excellent for this purpose, some problems remain with battery exchange, disconnection of the electrode, and infection of the implanted portion [8]. Moreover, with increasing use of medical imaging, an increasing problem has been that patients with artificial pacemakers cannot undergo magnetic resonance imaging and may have significant artifacts on computed tomography [29]. For these reasons, alternatives to the artificial pacemaker have been long awaited.

J.A.E. Spaan et al. (Eds.): Biopacemaking, BIOMED, pp. 149–167, 2007.
springerlink.com

Recent advancements in regenerative medicine have enabled cardiomyocytes to be regenerated from various stem cells such as embryonic stem cells and bone marrow mesenchymal stem cells [1, 3]. We previously reported that cardiomyocytes can be regenerated from bone marrow mesenchymal stem cells by exposing them to 5-azacytidine [23]. We have also reported that cardiomyocytes can be efficiently obtained from murine embryonic stem cells by stimulating them with Noggin, an inhibitor of bone morphogenetic proteins [38]. These regenerated cardiomyocytes have automaticity and show spontaneous beating. Others have reported that transplantation of primary cultured fetal or neonatal cardiomyocytes into an infarcted area improves cardiac function [39]. Furthermore, the regenerated cardiomyocytes can be purified from a mixture of stem cell-derived cells [16, 25]. It has been suggested that patients with bradyarrhythmias could be treated by transplantation of regenerated cardiomyocytes [20, 36]. The use of autologous bone marrow mesenchymal stem cells shows particular promise for clinical application, because the absence of immunorejection means that immunodepressant drugs are not required.

Action potentials in cardiomyocytes are regulated by a variety of ion channels. The expression of these ion channels and the shape of the action potentials varies with the location of the cardiomyocytes and their developmental stage [4, 24]. Before regenerated cardiomyocytes can be used for treatment of bradyarrhythmias, it needs to be determined what types of ion channels they express, whether they express adrenergic receptors [6, 9], and whether they express connexin43 and can make gap junctions to the adjacent host cardiomyocytes. The present study focused on these issues and investigated whether bone marrow mesenchymal stem cell-derived cardiomyocytes could be used as bio-pacemakers. We describe here the current status of the application of bone marrow mesenchymal stem cell-derived cardiomyocytes as bio-pacemakers, and discuss the remaining issues that need to be resolved.

2 Materials and Method

2.1 Cell Culture and Cardiomyocyte Induction

The bone marrow mesenchymal stem cells used were cardiomyogenic (CMG) cells, prepared as described previously [23]. Cells were cultured on 60 mm dishes in Iscove's modified Dulbecco's medium (IMDM; GIB-CO/Invitrogen, Carlsbad, CA, USA). To induce differentiation, cells were treated with 3 μmol/l of 5-azacytidine for 24 h and then maintained for several weeks. After 5-azacytidine, cardiomyocytes comprised approximately 20% of the cells, as described previously. The CMG cardiomyocytes began to spontaneously contract 2–3 weeks after 5-azacytidine treatment.

2.2 Action Potential Recording

Electrophysiological studies were performed in IMDM containing (in mmol/l) 1.49 $CaCl_2$, 4.23 KCl, and 25 HEPES (pH 7.4). Cultured cells were placed on the stage of an inverted phase contrast microscope (Diaphoto-300; Nikon) at 23°C. Action potentials were recorded using conventional glass microelectrodes filled with 3M KCl (DC resistance 15–30 MΩ). Membrane potentials were measured in current clamp mode (MEZ-8300 amplifier; Nihon Kohden, Tokyo, Japan) with the built-in 4-pole Bessel filter set at 1 kHz.

2.3 RNA Extraction and Reverse Transcription Polymerase Chain Reaction (RT-PCR) Analysis

Total RNA was extracted from cells isolated at 0, 2, 4 and 6 weeks following cardiomyocyte induction, using Trizol Reagent (GIBCO) as described previously [13]. RT-PCR was performed to detect Kir2.1, Kir2.2, K_V1.2, K_V1.4, K_V1.5, K_V2.1, K_V4.2, K_V4.3, MERG, K_VLQT1, KCNE1, HCN2, HCN4, Ca_V1.2, GIRK1, GIRK2B, GIRK4, Kir6.2, SUR2A, and GAPDH mRNA. The expression of β_1 and β_2-adrenergic receptor mRNAs was also analyzed. PCR was performed for 30–35 cycles, with each cycle consisting of denaturation at 94°C, annealing at 54–60°C, and amplification at 72°C (each for 1 min). The primers and PCR cycles used are shown in Table 1. Primary cultured cardiomyocytes were used as positive controls. Prior to the quantitative analysis, the linear range of the PCR cycles was measured for each mRNA, and the appropriate number of PCR cycles was determined. GAPDH was used as an internal control for each sample.

2.4 cAMP Accumulation Assays

Cells were incubated in serumfree medium with 10^{-4}mol/l of 3-isobutyl-1-methylxanthine (Sigma-Aldrich, St Louis, MO, USA) for 30 min, and stimulated with isoproterenol (Sigma-Aldrich) for 10 min. In some experiments, cells were preincubated with propranolol (Sigma-Aldrich) for 20 min. The medium was aspirated rapidly, and incubation was terminated by addition of 1 ml of ice-cold 0.1 N HCl. The lysates were centrifuged at 3,000 rpm for 10 min, and the supernatants were used as samples. The cAMP levels were counted by radioimmunoassay using an assay kit (YAMASA, Chiba, Japan).

2.5 Videotape Recording

The cultured cells were observed through an inverted phase contrast video microscope (IX70; Olympus, Tokyo, Japan) equipped with a 4×quartz objective lens and a 1×relay lens. The culture dishes were maintained at 33°C using a thermoplate. The cell images were obtained using an intensified charged couple device camera (CS220; Tokyo Denshi, Japan) and digitally videotaped with a VHS recorder (wv-DR9; Sony, Tpan). Beating rate was then measured using the recorded video images.

2.6 Construction and Transfection of Myosin Light Chain 2v-Promoted Enhanced Green fluorescent Protein (EGFP) Plasmid

An expression vector, pMLC2v-EGFP, was constructed by cloning a 2.7 kb *Hind*III–*Eco*RI fragment of the rat myosin light chain-2v (MLC-2v) promoter region [15, 26] into the *Hind*III–*Eco*RI site of pEGFP-1 (Clontech, Palo Alto, CA, USA), so that

Table 1. PCR primers used in this study

Ion channel	Subunit	Sense primers	Antisense primers
I_{K1}	Kir2.1	CGCCTTCATCATTGGTGCAG	GCCTCCACCATGCCTTCCAG
	Kir2.2	CGGGCTACGCTGTGTGACTG	ACCCTCCTCTGTGACCCTG
I_{to}	Kv1.2	CACCGGGAGACAGAGGGA	TCAGACATCAGTTAACAT
	Kv1.4	CCCACCACCATCATCAGACA	CATCAAAGCTGGGACGGTTC
	Kv1.5	CATCGGGAGACAGACCAC	TTACAAATCTGTTTCCCG
	Kv2.1	TACTTCTTCGACCGCCACCC	CAGGCAGTGTGTTGAGTGAC
	Kv4.2	GTGCATCTCGGCTTATGATGA	ACTCGGAGTGTGACAAAGGC
	Kv4.3	GGACAAGAACAAGCGGCA	CCACATGGTCTGACGGAAG
I_{Kr}	MERG	GCTGCCTGAGTATAAGCTGC	GTCCACGATGAGGTCCACTAC
I_{ks}	K_{vLQT1}	TTTGTTCATCCCCATCTCAG	GTTGCTGGGTAGGAAGAGC
	KCNE1	CTTGACGCCCAGGATGAGC	GGGGTTCACGACAATGGCTTCAG
$I_{K,ACH}$	GIRK1	AACTTCCATCGAAGCTGC	GGTCCTCCAGCCATCTTT
	GIRK4	GTGTTGAAAACCTTAGCGGC	CACCCTCTTCATCCTTCTCG
$I_{K,ATP}$	Kir6.1	TTCAGCCGCCATGCTGTGA	GAGGGTCTGAATCAGGATGG
	Kir6.2	TCTACGACCTGGCTCCTAGTGA	CTGGAGAGATGCTAAACTTGGGC
	SUR2A/2B	ACACGCTCCGCTCCAGGCTG	GCCAGGCAGAACAGCTGTCT
I_f	HCN2	CCAGGAGAAGTACAAGCAAG	GAGTAGAGGCGACAGTAGGT
	HCN4	AGCGACTTCCGGTTTTACTGG	CCGTGCCGTTTTGTAGACCTC
$I_{Ca,L}$	α_{1C}	CTTAGTGGAGAAGTCGGGG	GGAGAGGCAGAGCGAGAAG
β Receptor	β 1	ACGCTCACCAACCTCTTCAT	AGGGGCACGTAGAAGGAGAC
	β2	CCTCATGTCGGTTATCGTCC	GGCACGTAGAAAGACACAATC
GAPDH		AGTATGACTCCACTCACGGCAA	TCTCGCTCCTGGAAGATGGT

EGFP was expressed under the control of the MLC-2v promoter, as described previously [14, 19]. This plasmid also contained the neomycin resistance gene to enable selection of permanently transfected clones. MLC-2v is specifically expressed in ventricular cardiomyocytes. The MLC2v-EGFP plasmid was transfected into CMG cells by lipofection. After 24 h, when the cells were about 20% confluent, a mixture containing 2 μg of plasmid DNA and 4 μL of LT1 TransIT Polyamine Transfection Reagent (Mirus Bio, Madison, WI, USA) in OPTI-MEM (Life Technologies, Gaithersburg, MD, USA) were added to each 35 mm culture dish. After selection with 1,000 μg/ml of G418 for 4 weeks, stably transfected colonies derived from single cells were cloned and pooled. EGFP fluorescence was observed under a fluorescence microscope (TMD300; Olympus).

2.7 Flow Cytometry and Cell Sorting

Flow cytometry and sorting of EGFP (+) cells was performed on a FACS Aria (Becton Dickinson, Cockeysville, MD, USA). Cells were analyzed by light forward and side scatter and for EGFP fluorescence through a 530 nm bandpass filter as they traversed the beam of an argon ion laser (488 nm, 100 mΩ). Non-transfected control cells were used to set the background fluorescence. Cell sorting at 500 cells per second was performed 3 days after 5-azacytidine exposure, when the fluorescence of EGFP (+) cells was above background levels.

2.8 Cell Transplantation

Experimental procedures and protocols were approved by the Animal Care and Use Committees of Keio University. Female severe combined immunodeficient mice (aged 12 weeks) were anesthetized initially with ether and placed on a warm pad maintained at 37°C. The trachea was cannulated with a polyethylene tube connected to a respirator (Shinano, Tokyo, Japan), with tidal volume set at 0.6 ml and rate set at

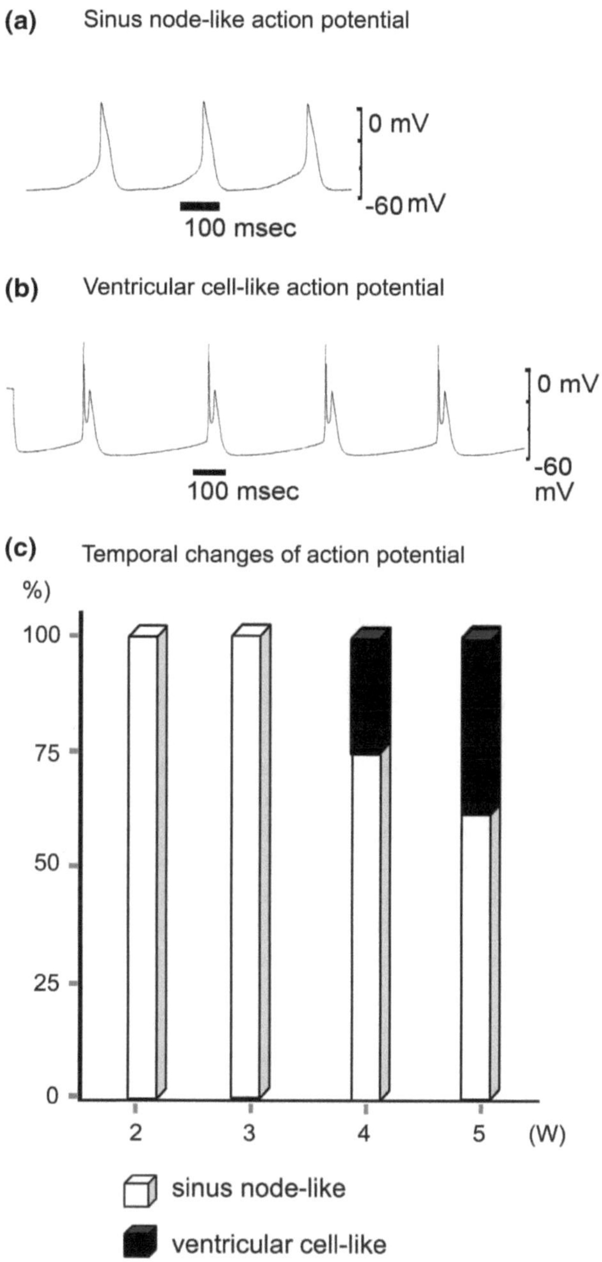

Fig. 1. Action potentials in CMG cardiomyocytes. Action potential recordings obtained from spontaneously beating cells at day 28 after 5-azacytidine treatment. Action potentials were categorized into 2 groups; sinus node-like action potentials **a**, and ventricular cardiomyocyte-like action potentials **b**. **c** All regenerated cardiomyocytes showed sinus node-like action potentials at first, but with time a progressively higher proportion showed ventricular cardiomyocyte-like action potentials (reprinted from [23]).

110 cycles/min. Mice were then anesthetized with isoflurane (0.5–1.5%) under controlled ventilation with the respirator for the remainder of the surgical procedure. A left thoracotomy was performed between ribs four and five, and the pericardial sac was removed. Isolated EGFP (+) cells that had been expanded for 5 days were resuspended in phosphate-buffered saline at a concentration of $5 \times \cdot 10^7$ cells/ml. A cell suspension volume of 50 μl was drawn into a 50 μl Hamilton syringe with a 31-gauge needle, and 10 μl was injected into the anterior wall of the left ventricle. Injection of phosphate-buffered saline was used as a control.

2.9 Statistical Analysis

Values are presented as mean ± SEM. The significance of differences between means was determined by ANOVA. Comparison of the control and treated groups was carried out using the non-parametric Fisher's multiple comparison tests. The accepted level of significance was $P < 0.05$.

3 Results

3.1 Temporal Changes of the Action Potentials in Bone Marrow-Derived Cardiomyocytes

To characterize the action potentials of the bone marrow-derived cardiomyocytes, we first recorded the action potentials of differentiated CMG cells at 2–5 weeks after treatment. Cardiomyocyte-like action potentials recorded from these spontaneously beating cells had (1) relatively long action potential durations or plateaus, (2) relatively shallow resting membrane potentials, and (3) pacemaker-like late diastolic slow depolarizations. Action potentials were distinctly sinus node-like (Fig. 1a) or ventricular myocyte-like (Fig. 1b). The sinus node-like action potentials had shallower resting membrane potentials and more marked late diastolic slow depolarizations, like pacemaker potentials. The ventricular myocyte-like action potentials had peak- and dome-like morphologies.

The two distinct types of action potential in CMG cells may reflect different developmental stages. Figure 1c shows how the proportions of sinus node-like and ventricular myocyte-like action potentials changed over successive weeks. Until 3 weeks, all of the action potentials recorded from CMG cells were sinus node-like. The ventricular myocyte-like action potentials were recorded from 4 weeks, and the percentage of these action potentials progressively increased there after. These findings suggested that bone marrow-derived cardiomyocytes showed temporal changes indicative of maturation.

3.2 Expression of K^+Cannels in CMG Cells

Cardiomyocytes have a wide variety of K^+ currents, including I_{K1},I_{to},I_{Kr},I_{Ks},$I_{K,Ach}$, and $I_{K,ATP}$. To begin to address the expression of inward rectifying K^+ (I_{K1}) channels, we first investigated expression of their constituting subunits Kir2.1 and Kir2.2 using

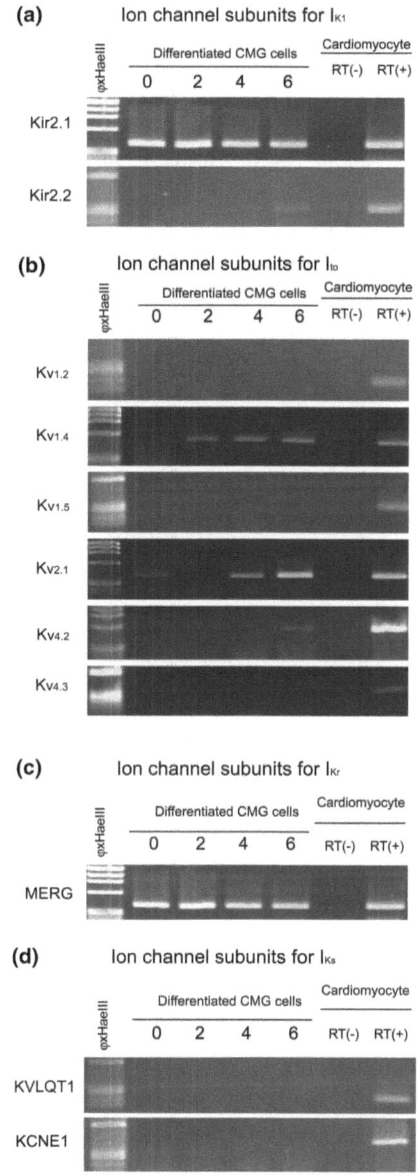

Fig. 2. Expression of I_{K1},I_{to},I_{Kr} and I_{Ks} ion channel subunits in CMG cells, shown by RT-PCR analysis of ion channel subunits. **a** The CMG cells expressed Kir2.1 before differentiation, and they also expressed Kir2.2 at 6 weeks when they showed ventricular myocyte-like action potentials. **b** K_v1.4 subunit was expressed from 2 weeks, and K_v2.1 subunit was expressed from 4 weeks. Expression of K_v4.2 was detected 6 weeks after differentiation, but K_v1.2, K_v1.5 and K_v4.3 were not detected before or any time after differentiation. **c** Mouse ERG (MERG) was expressed both before and after the differentiation. **d** CMG cells did not express K_vLQT1 or KCNE1 before or after differentiation.

RT-PCR. The expression of I_{K1} channels increases developmentally and the adult current density is achieved at postnatal days nine [32]. The basis of the developmental increase has been variously ascribed to increased channel number, open probability or unitary conductance [34]. We found that the CMG cells expressed Kir2.1 before differentiation, and that their expression was maintained (Fig. 2a). Kir2.2 was not expressed until 6 weeks, when they showed ventricular myocyte-like action potentials.

We next investigated expression of the transient outward K^+I_{to}) channel subunits $K_V1.2$, $K_V1.4$, $K_V1.5$, $K_V.2.1$, $K_V4.2$ and $K_V4.3$. Wickenden reported that during cardiomyocyte development, the predominant K^+ channel mRNA species switches from $K_V1.4$ to $K_V4.2$ and $K_V4.3$ [35]. Undifferentiated CMG cells expressed none of the subunits that compose I_{to} (Fig. 2b). Interestingly, the $K_V1.4$ subunit was expressed from 2 weeks, and the $K_V2.1$ subunit was expressed from 4 weeks. Expression of $K_V4.2$ was detected 6 weeks after differentiation, when the CMG cells showed ventricular myocyte-like potentials. However, neither $K_V1.2$, $K_V1.5$ nor $K_V4.3$ were induced in differentiated CMG cells. These expression patterns could explain how CMG cells developed ventricular cardiomyocyte-like action potentials at 4 or 6 weeks. We propose that an increase in the density of I_{to}, along with the previously reported $K_V1.4$ to $K_V4.2/K_V4.3$ isoform switch, produced the shortening of the action potential duration and the notchshaped transient repolarization following the spike (phase 1 of action potential).

Next, we investigated the expression of the ERG K^+ channels that underlie native rapidly activating delayed rectifier K^+ currents (I_{kr}), and the K_VLQT1 and KCNE1 channels that underlie the native slowly activating delayed rectifier K^+ current (I_{Ks}). Mouse ERG (MERG) was expressed both before and after differentiation (Fig. 2c), whereas K_VLQT1 and KCNE1 were not expressed either before or after differentiation (Fig. 2d).

The K^+ channel rapidly activated by application of acetylcholine ($I_{K,Ach}$ channel) is encoded by the combination of GIRK1, GIRK2B and GIRK4. We previously reported that bone marrow-derived cardio-myocytes express functional Ach receptors [13]. RT-PCR revealed that none of these subunits were expressed before cardiomyocyte induction (Fig. 3a). In contrast, the differentiated CMG cells expressed GIRK1 and GIRK2B after 6 weeks, when the cells showed ventricular myocyte-like action potentials.

ATP-sensitive inwardly rectifying K^+($I_{K,ATP}$) channels are expressed in a variety of tissues including heart. The $I_{K,ATP}$ channel has been reconstituted in functional form by coexpression of SUR, the sulfonylurea-binding protein, and the inwardly rectifying K^+ channel subunit, Kir6.2 [7]. CMG cells did not express these subunits before differentiation, but the differentiated cells expressed both Kir6.2 and SUR2A from 4 weeks onwards (Fig. 3b).

3.3 Expression of Pacemaker and Ca^{2+} Channels in CMG Cells

The hyperpolarization-activated, cyclic nucleotide-modulated (HCN) channel gene family is known to contribute significantly to cardiac pacemaker current (I_f). The present study demonstrated that the HCN family was not expressed before cardiomyocyte

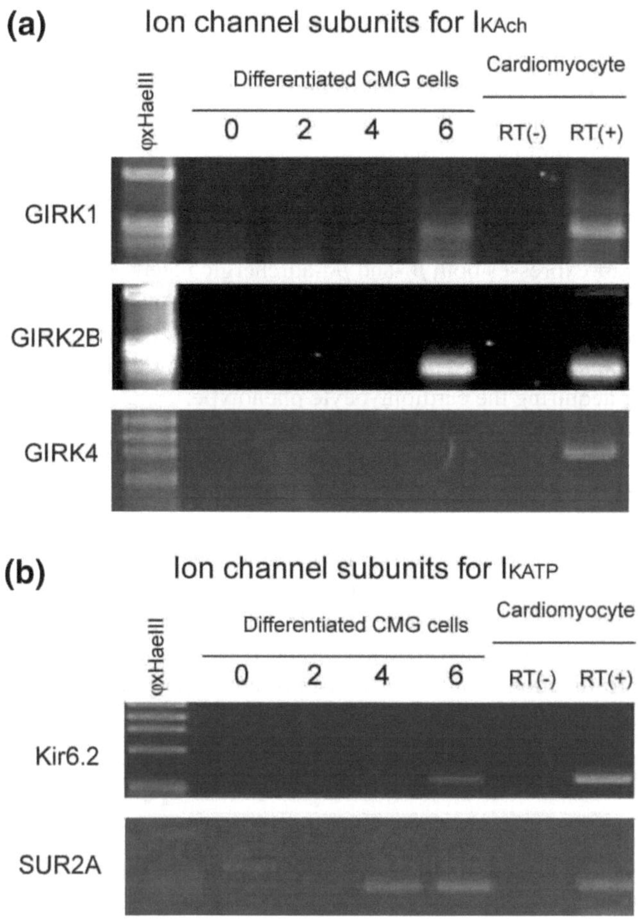

Fig. 3. Expression of $I_{K,Ach}$ and $I_{K,ATP}$ ion channel subunits in CMG cells. **a** The CMG cells expressed both GIRK1 and GIRK2B at 6 weeks. They did not express GIRK4. **b** Kir6.2 and SUR2A were expressed in CMG cells from 4 weeks after Differentiation.

differentiation, and that HCN4 and HCN2 was expressed from 2 to 6 weeks, respectively (Fig. 4a). These patterns of HCN channel expression in differentiated CMG cells were consistent with the developmental changes of normal cardiomyocytes.

The L-type Ca^{2+} channel in cardiomyocytes is composed of 1 α1, β and α2γ subunits. RT-PCR demonstrated that the α 1 subunit of the cardiac L-type Ca^{2+}channel (Ca_v1.2) was not expressed before CMG cell differentiation, but was expressed from 2 weeks (Fig. 4b), consistent with the finding of the action potential in differentiated cardiomyocytes.

We next examined whether the regenerated cardiomyocytes expressed functionally active cardiac L-type Ca^{2+} channels. Cardiac L-type Ca^{2+} channels are sensitive to the Ca^{2+}channel blocker verapamil. We recorded action potentials from the bone marrow-derived cardiomyocytes, added verapamil into the conditioned media, and observed the changes (Fig. 4c). Administration of verapamil decreased the beating rate of the bone marrow-derived cardiomyocytes, and shortened the duration and plateau of the action potential. These findings demonstrated that these cells expressed functionally active L-type Ca^{2+} hannels. Expression of the various ion channels during the differentiation process is summarized in Fig. 5.

3.4 Expression and Function of β_1-and β_2-Adrenergic Receptors in CMG Cells

Mammalian heart expresses both β_1-and β_2-adrenergic receptors, the β_1 receptor being the predominant sub-type (approximately 75–80% of total b receptors). CMG cells did not express β_1 and β_2 receptor transcripts before 5-azacytidine exposure, but the mRNAs were expressed from 1 week (Fig. 6a), indicating that they expressed β_1 and β_2 mRNA after acquiring the cardiomyocyte phenotype.

Isoproterenol, a β stimulant, increased the cAMP content of the CMG cells, and propranolol completely inhibited the isoproterenol-induced cAMP accumulation (Fig. 6b). Isoproterenol was applied to the cells to determine whether it would increase the spontaneous beating rate, and the results showed that it increased it significantly to 48% over the rate in the control cells. Preincubation with propranolol (a nonselective b blocker) or CGP20712A (a β_1-selective blocker) significantly suppressed the isoproterenol-induced increase in beating rate (by 78 and 71%, respectively; Fig. 6c). Taken together, these results indicated that the β_1 and β_2-adrenergic receptors expressed in CMG cells were functional, and that the isoproterenol-induced increase in spontaneous beating rate was mainly mediated by β_1 receptors.

3.5 Purification and Transplantation of Bone Marrow-Derived Cells into the Heart

Because our method produced not only cardiomyocytes but other cells such as adipocytes or osteoblasts, we needed to purify the cardiomyocytes prior to transplantation into the recipient heart. As shown in Fig. 7a–d, The MLC-2v-EGFP plasmid-transfected CMG cells showed EGFP signals after cardiomyocyte differentiation. After cell sorting using the EGFP signals, more than 99% of the sorted cells showed EGFP-positive signals (Fig. 7e–h).

Purified cells were then transplanted into the free walls of recipient left ventricles, and the success of transplantation was analyzed histologically. Mice transplanted with EGFP-positive cells were sacrificed at 2, 4, 8, and 12 weeks. EGFP-positive cardiomyocytes survived in the recipient hearts for more that 12 weeks (Fig. 8a–c). The orientation of the transplanted cells was parallel to the surrounding recipient cardiomyocytes. The control experiment revealed no EGFP-positive cells in the heart.

Next, we marked the purified cells with adenovirusmediated LacZ gene, and transplanted them into recipient hearts as above. Figure 8d shows the entire murine heart stained with LacZ at 4 weeks after transplantation, and Fig. 8e shows an enlargement of the site of injection. LacZ-positive regenerated cardiomyocytes were observed on the epicardial surface. Double staining with LacZ and hematoxylineosin

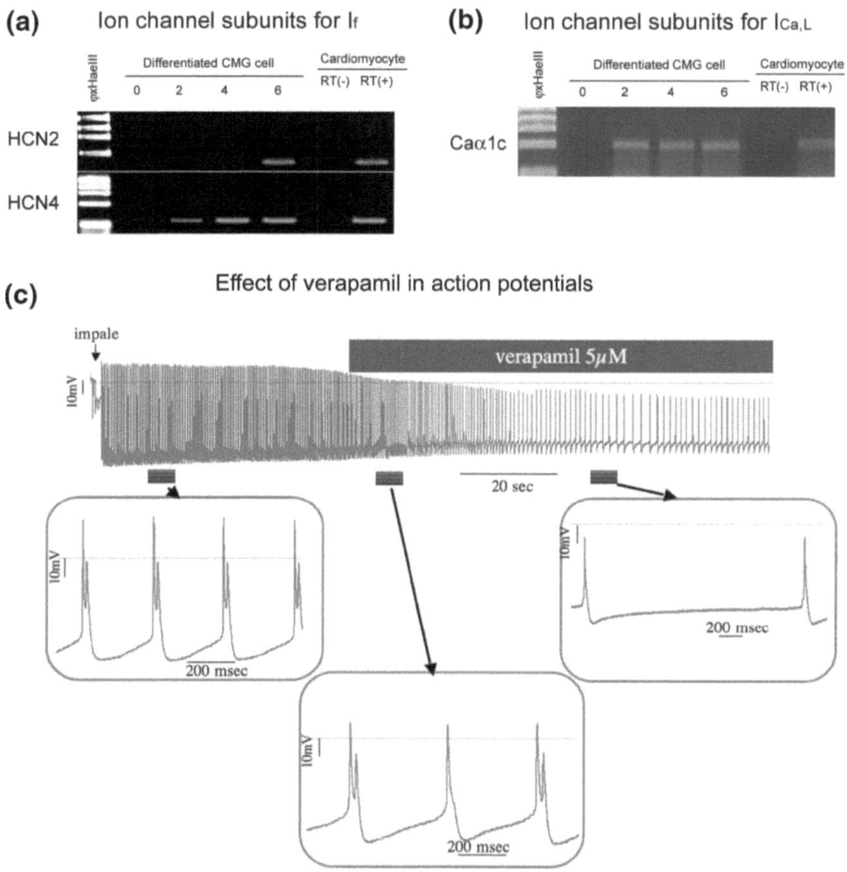

Fig. 4. Expression of pacemaker and Ca^{2+} channels in CMG cells. **a** The regenerated cardiomyocytes expressed HCN4 and HCN2 from 2 to 6 weeks after differentiation, respectively. **b** RT-PCR demonstrated that the α1 subunit of the cardiac L-type Ca^{2+} channel ($Ca_v1.2$) was expressed from 2 weeks after differentiation. **c** Treatment with verapamil decreased the beating rate of the bone marrow-derived cardiomyocytes, and the duration and plateau of the action potential shortened.

was performed on the transverse section of the transplanted heart at the midventricular level. The LacZ-stained transplanted cells were clearly distinguishable from the recipient cardiomyocytes, and were distributed as if they were patchwork into the surrounding tissue (Fig. 8f). Transverse and longitudinal sections of the transplanted cardiomyocytes at higher magnifications are shown in Fig. 8g, h. The transplanted cells were arranged in parallel with the recipient cardiomyocytes.

Triple-immunostaining for GFP, Toto3 (nucleus), and connexin43 is shown in Fig. 8i, j. Connexin43 was clearly observed between the transplanted regenerated cardiomyocytes and the adjacent recipient cardiomyocytes. These findings indicated that the transplanted regenerated cardiomyocytes made gap junctions with the surrounding recipient cardiomyocytes, and resided stably in the heart for a long periods.

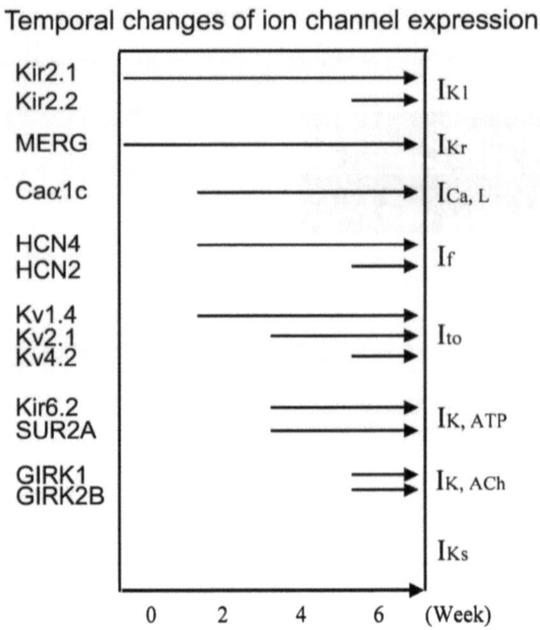

Fig. 5. Summary of the expression of various ion channels during the differentiation process in CMG cells

4 Discussion

The present study demonstrated that (1) bone marrow mesenchymal stem cells could differentiate into cardiomyocytes with automaticity, (2) these cells expressed β-adrenergic receptors and increased their beat rate in response to isoproterenol, and (3) the transplanted cardiomyocytes survived in the recipient heart and connected to the surrounding recipient cardiomyocytes via intercalated discs. We also found that the action potentials of the bone marrow-derived cardiomyocytes changed progressively with time. The temporal changes in action potentials occurred in parallel with the serial expression of various ion channels. Ion channels mediating I_{K1} and I_{Kr} were expressed prior to cardiomyocyte differentiation, suggesting that they were functional in the mesenchymal stem cells themselves or in both cardiomyocyte and nonmyocyte derivatives. During differentiation, the ion channels responsible for $I_{Ca,L}$ and If were expressed first, followed by the ion channels that mediate $I_{to}$and $I_{K,ATP}$. These findings strongly suggest that the differential expression of ion channels produced the temporal changes in action potentials.

Interestingly, bone marrow-derived cardiomyocytes expressed the pacemaker ion channels HCN4 after 2 weeks and HCN2 after 6 weeks. Moreover, they showed automaticity from 2 weeks, suggesting that HCN4 might be involved in the induction of automaticity. These findings suggest that they might be used as a bio-pacemaker when transplanted [28, 31]. The present study did not attempt to make sinoatrial or atrioventricular blocks, because the mice were too small for this to be practicable.

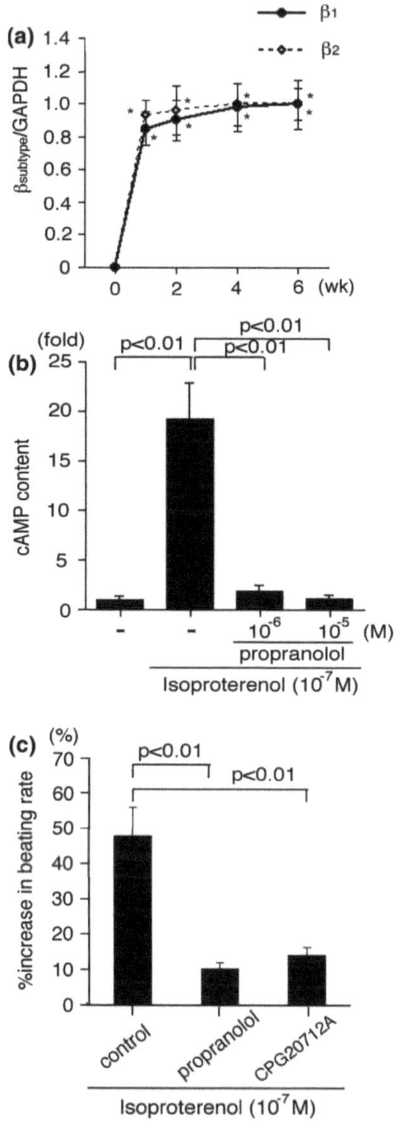

Fig. 6. β1-and β2-adrenergic receptor expression and signal transduction in CMG cells. **a** Quantitative analysis of the expression of β1-and β2-adrenergic receptors. Their expression was normalized relative to GAPDH. *P < 0.01 versus controls. **b** Cells were preincubated with propranolol (10^{-6} or 10^{-5} mol/l) for 20 min and stimulated with isoproterenol (10^{-7} mol/l) for 10 min. **c** Isoproterenol increased the spontaneous beating rate of CMG cells mainly via β1 receptors. CMG cells at 4 weeks after 5-azacytidine treatment were initially exposed to prazosin (10^{-6} mol/l) for 30 min to block β1-adrenergic receptors. Cells were then preincubated for 20 min with vehicle (phosphate-buffered saline), propranolol, or CGP20712A, and then stimulated with isoproterenol. The beating rate was counted 3 min after stimulation. Phosphate-buffered saline was added to the control.

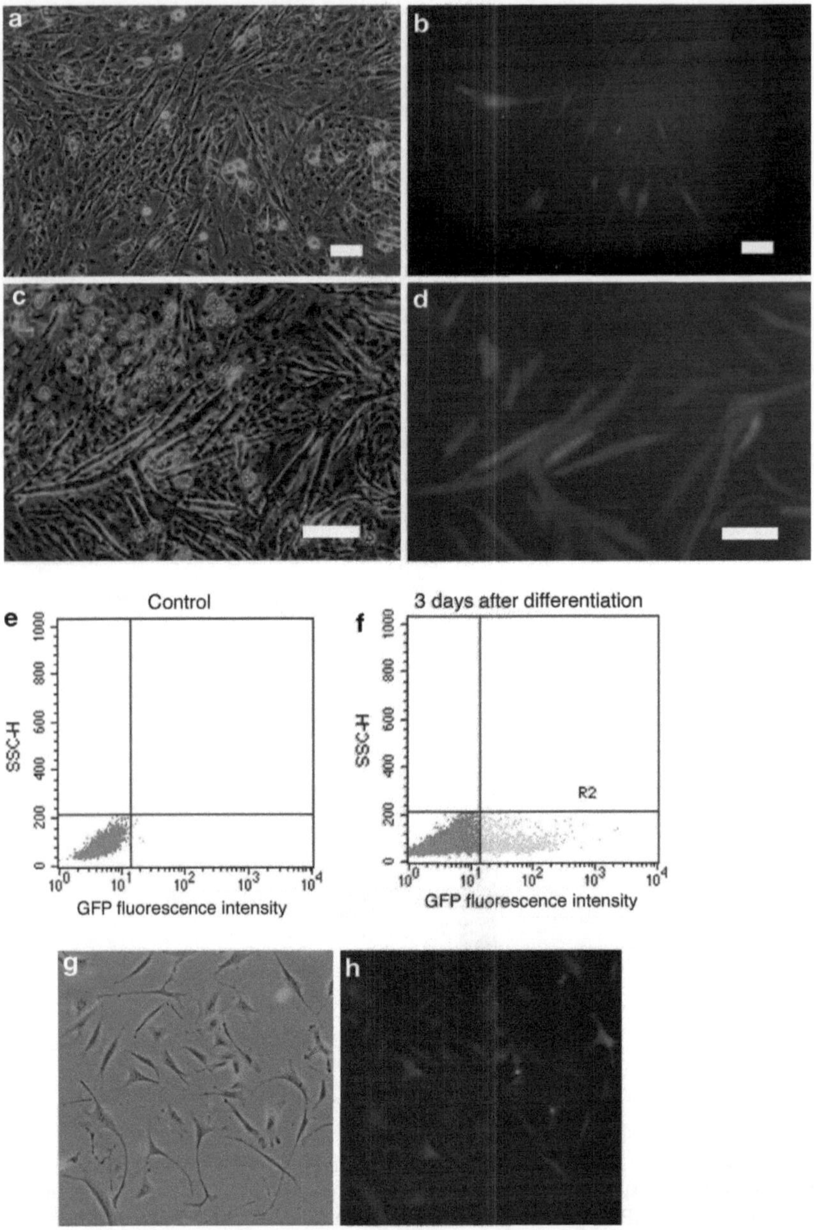

Fig. 7. Isolation of regenerated cardiomyocytes. Bone marrow mesenchymal stem cells were transfected with EGFP under the control of the promoter of a ventricular-myocardium-specific protein, myosin light chain-2v, and induced to differentiate. Some of the cells were GFP-positive 7 days after differentiation (**a**, **b**), and they started beating at 3 weeks (**c**, **d**). When the cells were fractionated with a FACS cell sorter after becoming GFP-positive (**e**, **f**), a highly purified preparation of cardiomyocytes was obtained. **g**, **h** Four days after cell sorting, all the cells were EGFP positive. (Reprinted and partly modified from [14]).

Moreover, the beating rate of the bone marrow-derived cardiomyocytes was approximately 80–160 beats/min under culture conditions, while the heart rate of the mice at rest ranged between 350 and 500 beats/min. Thus, in the present study, the transplanted cardiomyocytes could not have overcome the recipient heart rate. Possible reasons for the slow beat rate of the bone marrow-derived cardiomyocytes could include (1) the difference between the temperature in the cultured dish and body temperature [12], (2) the absence of humoral factors that affect heart rate, (3) the lack of sympathetic innervation [5], (4) the developmental stage [17], and (5) the part of the heart represented by the differentiation [33].

We believe that the successful application of mesenchymal stem cell-derived cardiomyocytes as bio-pacemakers will require a means to control their rate of automaticity (i.e., to augment their spontaneous beating rate). This might, for example, be accomplished by transfection of the ion channels mediating I_f (HCN2 and HCN4), and $I_{Ca,T}$ (CaV3.1–3.3), which play a crucial role in the pacemaker potential. Recently, Kuwahara et al. [22] showed that neuron-restrictive silencer factor (NRSF), a transcriptional repressor, selectively regulates expression of multiple fetal cardiac genes. It regulates expression of genes for atrial natriuretic peptide, brain natriuretic peptide, and α-skeletal actin, and plays a role in the molecular pathways leading to the re-expression of these genes in ventricular myocytes. Sequence analysis of the HCN4 gene by Kuratomi et al. [21] revealed the presence of a conserved neuron-restrictive silencer element (NRSE) motif, which is known to bind NRSF. These findings implicate the NRSE–NRSF system in HCN4 expression in cardiac myocytes. We therefore suggest that overexpression of NSRF in mesenchymal stem cell-derived cardiomyocytes might increase the spontaneous beating rate. Further studies will be required to determine whether this constitutes a possible method for controlling automaticity in regenerated cardiomyocytes.

Yasui et al. [37] studied action potentials and the occurrence of one of the pacemaker currents, I_f, by the whole-cell voltage- and current-clamp technique at the stage when a regular heartbeat is first established (9.5 days postcoitus) and at 1 day before birth. They showed a prominent I_f in mouse embryonic ventricles at the early stage, which decreased by 82% before birth in tandem with the loss of regular spontaneous activity by the ventricular cells. They concluded that I_f current of the sinus node type is present in early embryonic mouse ventricular cells. Loss of the I_f current during the second half of embryonic development is associated with a tendency for the ventricle to lose pace-maker potency. Our findings in bone marrow mesenchymal stem cell-derived cardiomyocytes may reflect the developmental changes in action potentials that occur in embryonic ventricular cardiomyocytes.

In the present study, we used mice for the experiments, which facilitated the molecular analysis. However, it is very difficult to make atrioventricular block or transplant cells into the atrioventricular node in mice. Therefore, we could not perform electrophysiological or functional analyses using this system. Further experiments would need to use large animals such as dogs or miniature pigs to enable more precise physiological analyses. Molecular analysis is difficult in large animals, but transplantation of regenerated cardiomyocytes using large animals would be required for preclinical studies before beginning clinical trials.

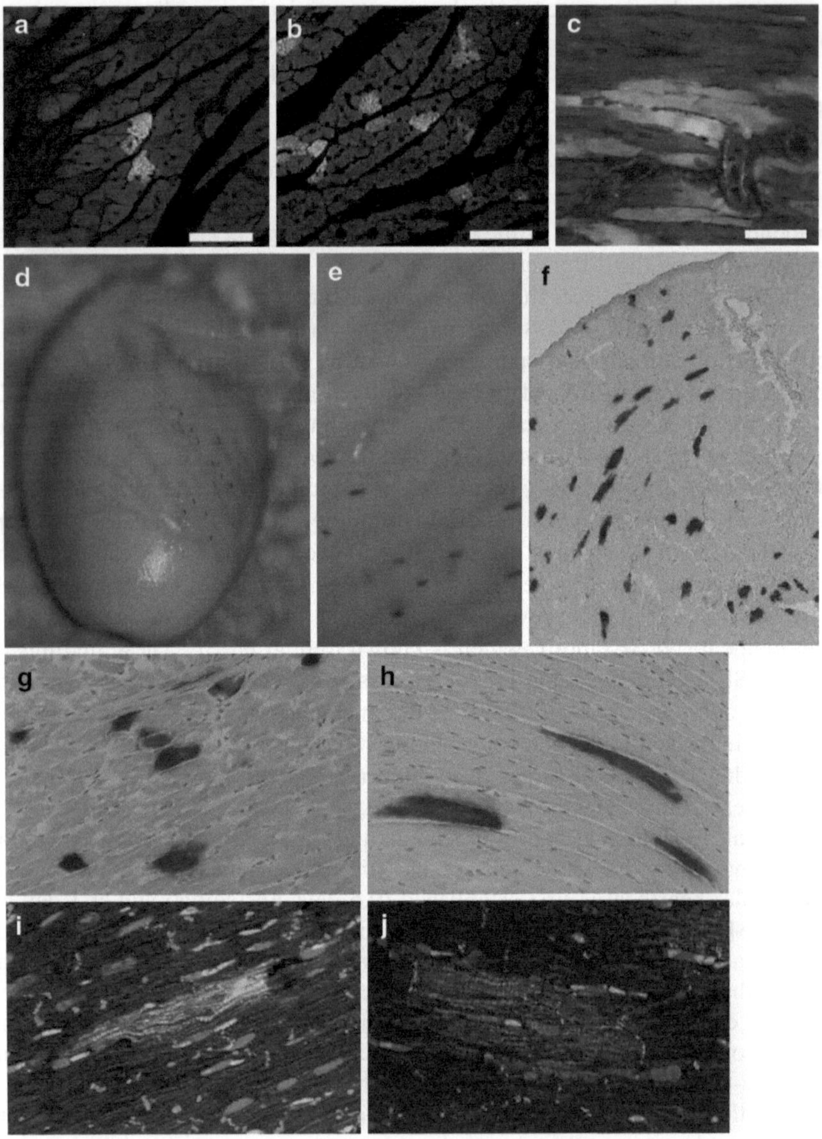

Fig. 8. Transplantation of regenerated cardiomyocytes. Regenerated cardiomyocytes were transplanted into the adult mice heart by injection. The transplanted cardiomyocytes showed stable long-term survival. After injection the regenerated cardiomyocytes adhered in small clusters to surrounding cardiomyocytes, where they assumed the form of mature cardiomyocytes that had the appearance of short strips of paper. **a–c** Cells in which the GFP fluorescence was observed; **d–h** cells transfected with the LacZ gene and stained. **i, j** Cells co-immunostained for connexin43 formed apparent gap junctions with the surrounding cardiomyocytes. EGFP appears green, nuclear staining with TOTO-3 is blue, and connexin43 labeling is red. (Reprinted and modified from [14]).

In the present study we purified the cardiomyocytes from the mixture of various types of cells prior to transplantation. Although recent advances in regenerative medicine and developmental biology have improved the efficiency of cardiomyocyte induction from various stem cells, it is at present impossible to selectively induce a specific type of cell. Bone marrow mesenchymal stem cells can differentiate into osteoblasts, chondroblasts, and adipocytes in addition to cardiomyocytes [2, 10, 11, 27, 30]. Thus, the purification of cardiomyocytes remains an indispensable step for optimization of tissue strength and efficient conduction of action potentials. However, the results of the cell transplantation by needle injection showed that the bone marrow mesenchymal stem cell-derived cells were scattered around the injected site, and formed a patchwork in the surrounding tissue. When relatively few donor cells are surrounded by a large number of host cells, the host cells may electrically clamp the donor cells and prevent their spontaneous activity. To solve this problem, it would be best to transplant the cells in clusters. We reported recently that fibrin polymercoated dishes can be used to produce cardiomyocyte cell sheets [18]. With further development, this system should allow transplantation of clumps of cardiomyocytes into the host heart.

Because bone marrow mesenchymal stem cells are autologous, the transplanted cells avoid immunorejection. This is a great advantage in using mesenchymal stem cells, but at the same time there are some drawbacks. The growth rate of mesenchymal stem cells is slower than for embryonic stem cells. However, the small size of the sinoatrial and atrioventricular nodes means that relatively few regenerated cardiomyocytes would be required for their repair. Bone marrow mesenchymal stem cells might therefore be a good resource for the repair of these tissues.

Acknowledgments. This study was supported by the program for Promotion of Fundamental Studies in Health Science of the National Institute of Biomedical Innovation (NIBIO), Japan, and research grants from the Ministry of Education, Science and Culture, Japan.

References

1. Amado LC et al (2005) Cardiac repair with intramyocardial injection of allogeneic mesenchymal stem cells after myocardial infarction. Proc Natl Acad Sci USA 102:11474– 11479
2. Ashton BA et al (1980) Formation of bone and cartilage by marrow stromal cells in diffusion chambers in vivo. Clin Orthop Relat Res 151:294–307
3. Boheler KR et al (2002) Differentiation of pluripotent embryonic stem cells into cardiomyocytes. Circ Res 91:189– 201
4. Brillantes AM, Bezprozvannaya S, Marks AR (1994) Developmental and tissue-specific regulation of rabbit skeletal and cardiac muscle calcium channels involved in excitation-contraction coupling. Circ Res 75:503–510
5. Choate JK, Feldman R (2003) Neuronal control of heart rate in isolated mouse atria. Am J Physiol Heart Circ Physiol 285:H1340–H1346
6. Choi YH et al (2006) Cardiac conduction through engineered tissue. Am J Pathol 169: 72–85

7. Chutkow WA et al (1996) Cloning, tissue expression, and chromosomal localization of SUR2, the putative drug-binding subunit of cardiac, skeletal muscle, and vascular KATP channels. Diabetes 45:1439–1445
8. Eberhardt F et al (2005) Long term complications in single and dual chamber pacing are influenced by surgical experience and patient morbidity. Heart 91:500–506
9. Edelberg JM, Aird WC, Rosenberg RD (1998) Enhancement of murine cardiac chronotropy by the molecular transfer of the human beta2 adrenergic receptor cDNA. J Clin Invest 101:337–343
10. Ferrari G et al (1998) Muscle regeneration by bone marrow-derived myogenic progenitors. Science 279:1528–1530
11. Friedenstein AJ, Chailakhyan RK, Gerasimov UV (1987) Bone marrow osteogenic stem cells: in vitro cultivation and transplantation in diffusion chambers. Cell Tissue Kinet 20:263–272
12. Geiser F, Baudinette RV, McMurchie EJ (1989) The effect of temperature on isolated perfused hearts of heterothermic marsupials. Comp Biochem Physiol A 93:331–335
13. Hakuno D et al (2002) Bone marrow-derived regenerated cardiomyocytes (CMG Cells) express functional adrenergic and muscarinic receptors. Circulation 105:380–386
14. Hattan N et al (2005) Purified cardiomyocytes from bone marrow mesenchymal stem cells produce stable intracardiac grafts in mice. Cardiovasc Res 65:334–344
15. Henderson SA et al (1989) Structure, organization, and expression of the rat cardiac myosin light chain-2 gene. Identification of a 250-base pair fragment which confers cardiacspecific expression. J Biol Chem 264:18142–18148
16. Hidaka K et al (2003) Chamber-specific differentiation of Nkx2.5-positive cardiac precursor cells from murine embryonic stem cells. FASEB J 17:740–742
17. Hou PC, Burggren WW (1989) Interaction of allometry and development in the mouse Mus musculus: heart rate and hematology. Respir Physiol 78:265–280
18. Itabashi Y et al (2005) A new method for manufacturing cardiac cell sheets using fibrincoated dishes and its electrophysiological studies by optical mapping. Artif Organs 29:95– 103
19. Kawada H et al (2004) Nonhematopoietic mesenchymal stem cells can be mobilized and differentiate into cardiomyocytes after myocardial infarction. Blood 104:3581–3587
20. Kehat I et al (2004) Electromechanical integration of cardiomyocytes derived from human embryonic stem cells. Nat Biotechnol 22:1282–1289
21. Kuratomi S et al (2006) NRSF regulates the developmental and hypertrophic changes of HCN4 transcription in rat cardiac myocytes. Biochem Biophys Res Commun 353:67–73
22. Kuwahara K et al (2003) NRSF regulates the fetal cardiac gene program and maintains normal cardiac structure and function. EMBO J 22:6310–6321
23. Makino S et al (1999) Cardiomyocytes can be generated from marrow stromal cells in vitro. J Clin Invest 103:697–705
24. Maltsev VA et al (1994) Cardiomyocytes differentiated in vitro from embryonic stem cells developmentally express cardiac-specific genes and ionic currents. Circ Res 75:233– 244
25. Muller M et al (2000) Selection of ventricular-like cardiomyocytes from ES cells in vitro. FASEB J 14:2540–2548
26. O'Brien TX, Lee KJ, Chien KR (1993) Positional specification of ventricular myosin light chain 2 expression in the primitive murine heart tube. Proc Natl Acad Sci USA 90:5157–5161
27. Rickard DJ et al (1994) Induction of rapid osteoblast differentiation in rat bone marrow stromal cell cultures by dexamethasone and BMP-2. Dev Biol 161:218–228

28. Rosen MR et al (2004) Recreating the biological pacemaker. Anat Rec A Discov Mol Cell Evol Biol 280:1046–1052
29. Rozner MA, Burton AW, Kumar A (2005) Pacemaker complication during magnetic resonance imaging. J Am Coll Cardiol 45:161–162 (author reply 162)
30. Umezawa A et al (1992) Multipotent marrow stromal cell line is able to induce hematopoiesis in vivo. J Cell Physiol 151:197–205
31. Viswanathan PC et al (2006) Recreatingan artificial biological pacemaker: insights from a theoretical model. Heart Rhythm 3:824–831
32. Wahler G.M (1992) Developmental increases in the inwardly rectifying potassium current of rat ventricular myocytes. Am J Physiol 262:C1266–C1272
33. Wei H et al (2005) Embryonic stem cells and cardiomyocyte differentiation: phenotypic and molecular analyses. J Cell Mol Med 9:804–817
34. Wetzel GT, Klitzner TS (1996) Developmental cardiac electrophysiology recent advances in cellular physiology. Cardiovasc Res 31:E52–E60
35. Wickenden AD et al (1997) Effects of development and thyroid hormone on K+ currents and K+ channel gene expression in rat ventricle. J Physiol 504:271–286
36. Xue T et al (2005) Functional integration of electrically active cardiac derivatives from genetically engineered human embryonic stem cells with quiescent recipient ventricular cardiomyocytes: insights into the development of cell-based pacemakers. Circulation 111:11–20
37. Yasui K et al (2001) I(f) current and spontaneous activity in mouse embryonic ventricular myocytes. Circ Res 88:536–542
38. Yuasa S et al (2005) Transient inhibition of BMP signaling by Noggin induces cardiomyocyte differentiation of mouse embryonic stem cells. Nat Biotechnol 23:607–611
39. Zimmermann WH et al (2006) Engineered heart tissue grafts improve systolic and diastolic function in infarcted rat hearts. Nat Med 12:452–458

Enrichment of Cardiac Pacemaker-Like Cells: Neure gulin-1 and Cyclic amp Increase I_f-Current Density and Connexin 40 mRNA Levels in Fetal Cardiomyocytes

Arjang Ruhparwar[1], Fikret Er[2], Ulrich Martin[3], Kristin Radke[3], Ina Gruh[3], Michael Niehaus[4], Matthias Karck[1], Axel Haverich[3], and Uta C. Hoppe[2,5]

[1] Department of Cardiac Surgery, University of Heidelberg, 69120 Heidelberg, Germany
Arjang.Ruhparwar@med.uni-heidelberg.de
[2] Department of Internal Medicine III, University of Cologne, Kerpener Str. 62, 50937 Cologne, Germany
[3] Department of Thoracic and Cardiovascular Surgery, Hannover Medical School, Carl-Neuberg-Str. 1, 30625 Hannover, Germany
[4] Department of Cardiology and Angiology, Hannover Medical School, Carl-Neuberg-Str 1, 30625 Hannover, Germany
[5] Center for Molecular Medicine, University of Cologne (CMMC), Cologne, Germany

Abstract. Generation of a large number of cells belonging to the cardiac pacemaker system would constitute an important step towards their utilization as a biological cardiac pacemaker system. The aim of the present study was to identify factors, which might induce transformation of a heterogenous population of fetal cardiomyocytes into cells with a pacemaker-like phenotype. Neuregulin-1 (α-and *β*-isoform) or the cAMP was added to fresh cell cultures of murine embryonic cardiomyocytes. Quantitative northern blot analysis and flowcytometry were performed to detect the expression of connexins 40, 43 and 45. Patch clamp recordings in the whole cell configuration were performed to determine current density of I_f, a characteristic ion current of pacemaker cells. Fetal cardiomyocytes without supplement of neuregulin or cAMP served as control group. Neuregulin and cAMP significantly increased mRNA levels of connexin 40 (Cx-40), a marker of the early differentiating conduction system in mice. On the protein level, flowcytometry revealed no significant differences between treated and untreated groups with regard to the expression of connexins 40, 43 and 45. Treatment with cAMP (11.2 ± 2.24 pA/pF; $P < 0.001$) and neuregulin-1-*β* (6.23 ± 1.07 pA/pF; $P < 0.001$) significantly increased the pacemaker current density compared to control cardiomyocytes (1.76 ± 0.49 pA/pF). Our results indicate that neuregulin-1 and cAMP possess the capacity to cause significant transformation of a mixed population of fetal cardiomyocytes into cardiac pacemaker-like cells as shown by electrophysiology and increase of Cx-40 mRNA. This method may allow the development of a biological cardiac pacemaker system when applied to adult or embryonic stem cells.

Keywords: Biological pacemaker, Transplantation, Conduction system.

1 Introduction

The heart is endowed with specialized excitatory and conducting myocytes that are responsible for the generation and conduction of rhythmical impulses and contractions throughout the heart [2]. When these cells are damaged by disease, the

J.A.E. Spaan et al. (Eds.): Biopacemaking, BIOMED, pp. 168–177, 2007.
springerlink.com

implantation of an electronic cardiac pacemaker becomes necessary [15]. Potential alternative approaches may derive from cell or gene therapy. In a first study we were able to show that transplanted fetal cardiomyocytes can function as a biological cardiac pacemaker [25]. Miake et al. [17] postulated latent pacemaker capability of working myocardium. This potential ability is inhibited by the inward-rectifier potassium current I_{K1} encoded by the gene Kir2 which is not expressed in pacemaker cells. Following dominant-negative suppression of I_{K1}, ventricular myocytes exhibited spontaneous activity with their action potentials resembling typical patterns of genuine pacemaker cells. Qu et al. [22] were able to show that adenovirus-mediated overexpression of the hyperpolarization-activated, cyclic nucleotidegated pacemaker current HCN2 provides an I_f-based pacemaker sufficient to drive the heart when injected into a localized region of the atrium. Finally, Potapova et al. [21] demonstrated that transplantation of HCN2-transfected human mesenchymal stem cells (hMSCs) leads to expression of functional HCN2 channels in vitro and in vivo, mimicking overexpression of HCN2 genes in cardiac myocytes.

Transplanting cardiomyocytes as cardiac pacemaker may open a new perspective for the treatment of cardiac arrhythmia such as sinus node dysfunction or atrioventricular block (AV-block), ranging from infants and premature babies with congenital AV-block (incidence: one in 20,000–25,000 live births) who might be too small for the treatment with artificial pacemakers, to patients with acquired disease [11, 18]. However, so far the number of transplantable cells of the cardiac conduction system (CCS), regardless of their origin (fetal, adult and stem cell derived) is limited. On the way to a biological cardiac pacemaker it is therefore imperative to enrich pacemaker-like cells and, thus, to develop a method enabling transformation of a heterogeneous population of cardiomyocytes into cells of the CCS. Recently, Gassanov et al. [10] demonstrated that endothelin-1 directed differentiation of embryonic stem cells towards a pacemaker phenotype. Moreover, neuregulin-1 was shown to induce ectopic expression of the lac-Z conduction marker in vivo in CCS-lac-Z reporter mice within a short period of 8.5–10.5 days post coitum (dpc) [16, 23]. This indicated that neuregulin-1 might promote formation of the murine CCS. However, so far no studies have been performed of any inductive factors on cultured murine embryonic cardiomyocytes, which might prove useful for cell transplantation.

Therefore, the aim of the present study was to possibly transform a heterogenous population of cultured mouse embryonic cardiomyocytes (13 dpc) into cells with pacemaker-like characteristics by the separate exposure to the α-and β-isoforms of neuregulin-1, and cAMP. To determine the extent of enrichment of cells with a pacemaker-like phenotype we evaluated the expression of connexin 40 (Cx-40), a marker of the early differentiating conduction system in mice [5, 19], and secondly recorded the hyperpolarization-activated inward current I_f, a characteristic ion current of pacemaker cells [7, 10]. Our results show a significant enhancement of Cx-40 mRNA expression and increase of I_f current density upon neuregulin and cAMP treatment, indicating direction of differentiation towards pacemaker-like cells in a mixed population of embryonic cardiomyocytes.

2 Methods

2.1 Mouse Embryonic Cardiomyocytes

These studies were approved by the Hannover Medical School Animal Care Committee and were conducted in accordance with guidelines of the Government of Lower Saxony.

Single cell suspensions were prepared by 0.1%-collagenase II-digestion (Worthington Biochemical Corp.) of hearts at 37°C for 30 min, harvested from 13-dpc embryos of NMRI-mice (Charles River Laboratories, Boston MA). The cells were plated onto gelatinecoated dishes. More than 90% of the cells were viable, as evidenced by dye exclusion assay.

2.2 Treatment with *α*-and *β*-Neuregulin-1

All cytokines were aliquoted, stored and transferred in siliconized eppendorf tubes and pipette tips. Cells were distributed in 10-mm petri dishes, each dish containing 8×10^6 cells. The petri dishes were divided into three groups: Group 1: cells were supplemented with 100 μg/ml neuregulin-1-α (Recombinant human HRG- α EGF domain, R&D-Systems; size 7 kDa; stock 1 mg/ml in 0.1% BSA/PBS) for 48 h. Group 2: cells were supplemented with 100 lg/ml neuregulin-1-*β* (Recombinant human NRG-1-*β*1/HRG- *β*1 EGF-domain, R&D-Systems; size 8 kDa; stock 1 mg/ml in 0.1% BSA/PBS) for 48 h. Group 3: control group without neuregulin-1. The medium consisted of DMEM, 10% FBS, 1% glutamin/penicillin/streptomycin and 1% non-essential amino acids. The medium was changed every 12 h, whereby neuregulin was added each time.

2.3 Treatment with cAMP

The petri dishes were divided into two groups: Group 1: cells were supplemented with 5 mmol/l db-cAMP (dissolved in a PBS vehicle) for 48 h. Group 2: control group without db-cAMP. The medium consisted of DMEM, 10% FBS, 1% glutamin/penicillin/streptomycin and 1% non-essential amino acids. The medium was changed every 12 h, whereby db-cAMP was added each time.

2.4 Electrophysiology

Experiments were carried out with the use of standard microelectrode whole-cell patch-clamp techniques with an Axopatch 200B amplifier (Axon Instruments) while sampling at 10 kHz and filtering at 2 kHz [9]. Current recordings were performed at room temperature (21– 23°C). The recording bath solution contained (in μmol/l) KCl 100, NaCl 35, $CaCl_2$ 2, glucose 10, $MgCl_2$ 1, HEPES 10; pH was adjusted to 7.4 with NaOH. $BaCl_2$ 2 μmol/l, $CdCl_2$ 200 μmol/l, and 4-aminopyri-dine 4 μmol/l were added to block I_{K1},I_{CaL} and I_{to}, respectively. The micropipette electrode solution was composed of (in μmol/l) K-glutamate 130, KCl 15, NaCl 5, $MgCl_2$ 1, HEPES 10, and Mg-ATP 5; pH was adjusted to 7.3 with KOH. Borosilicate microelectrodes had tip resistances of 3–5 MΩ when filled with the internal recording solution.

2.5 Northern Blot Analysis

Cells were homogenized in Trizol reagent (Gibco) and purified according to the manufacturers instructions. After phenol-chloroform extraction and sodiumace-tate precipitation RNA samples were quantitated by spectrophotometry at 260 nm. After seperation on an agarose gel (15 μg/lane) the RNA was transferred to a nylon membrane (Hybond-N, Amersham) and UV-crosslinked. After prehybridization for 1 h at 65°C the membrane was incubated with DIG-labeled (Boeh-ringer Mannheim DIG RNA labelling kits SP6/T7) oligonucleotides of Cx-40 (372 bp) or cardiac troponin-I (TnI, 195 bp) at 68°C over night. Amounts of the RNA samples were confirmed by additionally running each blot of Cx-40 and TnI with β-actin (304 bp) nucleotides. The signals were visualized using α-DIG-AP (1:10,000, Roche) in 1% block reagent and CDP-Star reagent (Boehringer Mannheim). Neonatal heart tissue and skeletal muscle served as positive and negative control group.

2.6 Quantification of Northern Blot Signals

The blots were scanned (Sharp JX-330). Quantification was performed by measuring the optical density (pixel and area) of the blot signals with the ''Enhanced analysis system'' for gel documentation (Herolab GmbH, Wiesloch, Germany). To normalize the amount of RNA/lane optical density values for Cx-40 and troponin I of different groups (group 1 treated with NRG-1-α, group 2 NRG-1-β, group 3 control group) were divided by the corresponding optical density of β -actin. To normalize variability in the proportion of cardiomyocytes in different cell samples, the resulting values for Cx-40 (Cx-40/β-actin) were divided by the values for cardiac specific troponin I (troponin I/β-actin). All experiments were repeated at least three times.

2.7 Flowcytometry

Due to small amounts of membrane protein flowcytometry rather than western blotting was performed to evaluate protein expression. For identification of cardiomyocytes, cells were stained with a mouse IgG1 monoclonal antibody EA-53 to sarcomeric α-actinin (diluted 1:800, Sigma) using the BD Cytofix/Cytoperm Kit (BD Biosciences, Heidelberg, Germany) according to the manufacturer's instructions. Co-staining was performed using the following primary antibodies: a rabbit polyclonal antibody to mouse Cx-40 (diluted 1:100), a rabbit polyclonal antibody to mouse Cx-43 (diluted 1:25), or a rabbit polyclonal antibody to Cx-45 (diluted 1:50), respectively (all connexin antibodies from Chemicon, Temecula, CA, USA).

After incubation with indodicarbocyanin (Cy5) labeled donkey anti mouse IgG (diluted 1:500, Dianova, Hamburg, Germany) and carbocyanin (Cy2) labeled goat anti rabbit IgG (1:100, Dianova) as secondary antibodies, α-actinin positive cardiomyocytes were analyzed for connexin expression by flowcytometry using a FACSCalibur instrument (BD Biosciences, Heidelberg, Germany). Data were further processed using WinMDI 2.8 software, MFI (Martz, Eric. 1992-2001. http://www.umass.edu/microbio/mfi), and GraphPad Prism (version 3.02 for Windows, GraphPad Software, San Diego, CA, USA). Cardiomyocytes incubated with isotype control antibodies or rabbit serum served as negative controls.

2.8 Statistical Analysis

Pooled data are presented as mean ± standard error of the mean (SEM). Comparisons between groups were performed using one-way ANOVA. *P*-values less than 0.05 were deemed significant.

3 Results

3.1 Neuregulin and cAMP Increase mRNA Levels of Connexin 40

To investigate whether neuregulin or cAMP might induce differentiation of embryonic cardiomyocytes towards cells with a pacemaker-like phenotype RNA levels of Cx-40, a marker of conduction cells, were analyzed by northern blot experiments. Embryonic cardiomyocytes, isolated from 13-dpc mouse embryos were treated with neuregulin-1-α, neuregulin-1-*β* and cAMP. Untreated cells served as negative control. Two blots with RNA of each aliquot were prepared simultaneously and were hybridized with riboprobes specific for Cx-40 or cardiac troponin I. After signal detection by autoradiography, both blots were hybridized with *β*-actin specific probe to allow nor-malization of the amounts of loaded RNA. The differences in Cx-40 expression were calculated by the densitometric Cx-40 value with reference to the amount of tested RNA (*β*-actin) divided by the cardiac troponin I signal which itself was adjusted to the RNA content to only detect myocardial RNA without any interference of fibroblasts (fibroblasts make up to 50–70% of myocardial cells). Mean data of repeated series show that the normalized values of Cx-40 expression were more than two-fold increased by both pretreatment with neuregulin α and *β* compared to the control group (Fig. 1a). Moreover, cAMP exposure resulted in a three-fold upregulation of Cx-40 mRNA levels compared to control values (Fig. 1b).

3.2 Protein Levels of Connexins Are Not Modified by Neuregulin or cAMP

Since changes of mRNA levels may not readily reflect alterations of membrane proteins we performed additional analysis of connexin expression levels by flowcytometry following immunostaining. Day 13 fetal cardiomyocytes were treated with neuregulin-1-α, neuregulin-1-*β* and cAMP whereas time-matched untreated cells served as negative control. To omit any interference by non-myocardial cells only α-actinin expressing cells were analyzed. In contrast to our northern blot experiments showing a clear increase of Cx-40 mRNA levels upon neuregulin or cAMP exposure, we did not obtain any effect by either substance on Cx-40 protein expression compared to control cells (Table 1). Moreover, there was no change of the protein levels of Cx-43 and Cx-45 indicating that also the relative expression levels of different connexin subtypes remained unaffected (Table 1).

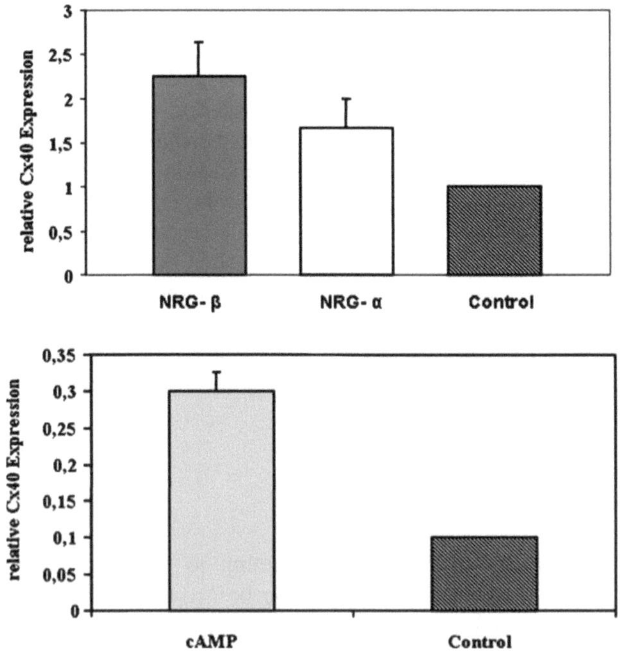

Fig. 1. a The degree of Cx-40 expression in neuregulin-1 treated cells and in the control group, evaluated by optical densitometry. Neuregulin-1 α has a significant positive impact on Cx-40 expression. **b** The degree of Cx-40 expression in cAMP treated cells and in the control group, evaluated by optical densitometry. cAMP has a significant positive impact on Cx-40 expression.

Table 1. Expression levels of connexins as identified by mean fluorescence intensity (MFI) of α-actinin positive cardiomyocytes after immunostaining and analysis by flowcytometry

	Mean MFI	SD	*N*	*P*
Connexin 40				
Control	29.4	8.8	5	NS
Camp	34.3	3.0	5	
Neuregulin α	32.3	9.9	3	
Connexin 43				
Control	61.3	9.6	3	NS
Camp	80.5	11.7	4	
Neuregulin α	57.6	9.0	3	
Connexin 45				
Control	537.7	84.7	3	NS
Camp	649.5	123.3	4	
Neuregulin α	564.5	71.4	2	

There was no significant impact of neuregulin or cAMP on the expression of Cx-40, Cx-43 and Cx-45

3.3 Neuregulin and cAMP Increase Pacemaker Current Size

Since the I_f current is characteristic for pacemaker cells, we evaluated the effect of neuregulin and cAMP on I_f current density of 13-dpc embryonic cardio-myocytes. After isolation cardiomyocytes were incubated in normal medium, medium with the addition of cAMP 5 mmol/l or medium with the addition of NRG-1-β 100 μg/ml for 48–60 h. Before voltage-clamp experiments monolayer cultures were dispersed by trypsin, and replated at a low density to study isolated cells within 2–8 h. To prevent any acute effects neuregulin and cAMP were washes out 10 min before performing patch-clamp experiments. Original current recordings and mean data demonstrate that compared to control cells (1.76 ± 0.49 pA/pF; 92 ± 3 pF; $n = 18$; at –130 mV) both neuregulin (6.23 ± 1.07 pA/pF; 94 ± 4 pF; $n = 12$; $P < 0.001$) and cAMP (11.2 ± 2.24 pA/pF; 98 ± 6 pF; $n = 8$; $P < 0.001$) significantly increased I_f current density without significantly affecting cell size (Fig. 2).

4 Discussion

Considering the feasibility of cell therapy aiming to replace dysfunctional cardiac tissue several obstacles remain to be overcome before this technology can enter any serious clinical practice. Selection of specific cell types for specific applications will be necessary. Strategies need to be developed for directing in vitro differentiation to a specific lineage. Thus, a first step is to identify and characterize candidate cells and to determine their developmental mechanisms. So far, methods to generate enriched cultures of cells with pacemaker-like properties have not been developed due to the lack of factors mediating CCS-specific differentiation and due to the lack of CCS-specific surface markers or CCS-specific promoters, which could allow specific selection [14, 20]. This report describes a simple way to enhance characteristics of pacemaker-like cells in embryonic cardiomyocytes by neuregulin or cAMP treatment.

Previously, neuregulin-mediated induction of differentiation of murine embryonic cells towards cells from the CCS has been suggested by in vivo observations of transgenic mice [23]. These data and our findings indicate an optimal time point for the intervention at day 9–13 [10, 23]. We now obtained enhancement of Cx-40 mRNA levels after neuregulin exposure in in vitro experiments of 13-dpc embryonic cardiomyocytes. Cx-40 has predominantly been detected in the conduction tissue of developing mammalian hearts and in the adult murine cardiac CCS and sinus node [1, 4, 5]. Interestingly, cAMP led to an even more pronounced increase of Cx-40 mRNA content compared to control conditions. It remains speculative, why the observed neuregulin- and cAMP-induced elevations of mRNA levels were not associated with any detectable increase of the absolute amount of Cx-40 proteins in the cell membrane. We also excluded any relative shift of expression levels of different connexin isoforms since immunostaining of Cx-43 and Cx-45 remained unchanged following neuregulin and cAMP treatment. Thus, it possibly requires longer time until an alteration of connexin membrane proteins becomes evident or the substances tested might exert inhibitory effects on posttranscriptional processing of Cx-40.

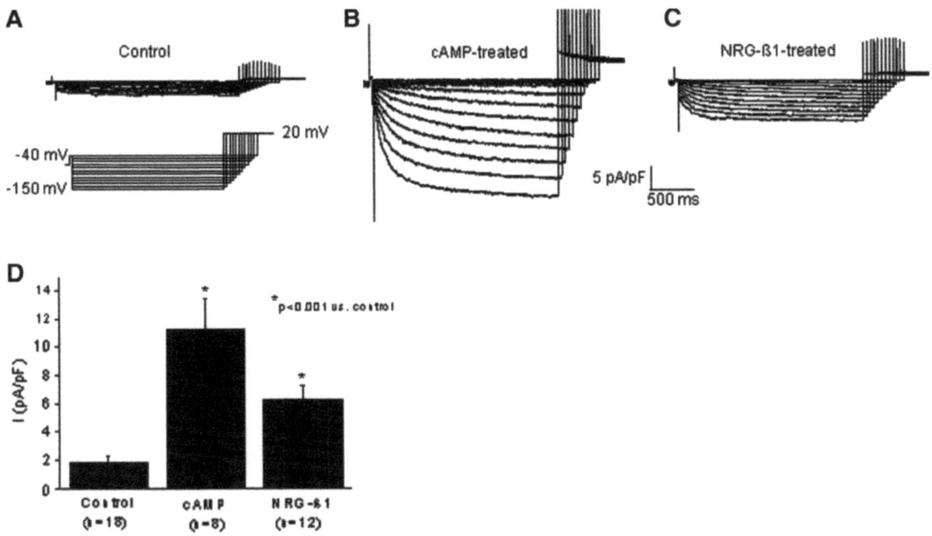

Fig. 2. Effect of cAMP and NRG-1-β treatment versus control on I_f current density (**a–d**). Original patch-clamp recordings in the whole cell configuration (**a–c**). Compared to control (a), cAMP (**b**) and NRG-1-β (**c**) significantly increased the I_f current density.

In addition to changes of Cx-40 mRNA levels we demonstrated a significant increase of I_f current density by neuregulin and cAMP. The pacemaker current I_f plays a central role for spontaneous diastolic depolarization of sinus node cells [26]. While small I_f currents may also be detected in working myocardium [3, 13], I_f current size is markedly larger in pacemaker cells [6]. Acute stimulation of I_f by cAMP is well known to shift the activation curve to more positive potentials and, thus, to increase current amplitude at half maximal activation [8, 12]. cAMP binds to the cytoplasmatic site of HCN channels which permits the channels to open more rapidly and completely after repolarization of the action potential with the result of accelerating rhythmogenesis [28]. However, maximal current size remains unaffected by acute sympathetic stimulation [8, 12]. In the present experiments we omitted any acute effects of both cAMP and neuregulin. In contrast to typical acute modulation of I_f we observed a significant increase of maximal current density at full activation potentials indicating elevated channel expression. Evaluation of expression differences of HCN isoforms will be done as a next step to verify the mechanism by which cAMP and neuregulin-1 influence I_f in embryonic cardiomyocytes.

Our present observations give further insight into the differentiation of the CCS. Moreover, our approach constitutes an elegant way to direct differentiation towards pacemaker-like cells out of a heterogeneous cell population. Generally, embryonic cardiomyocytes might not prove optimal candidates for cell transplantation due to low availability, their allogenic nature and possible ethical considerations. However, embryonic and fetal cardiomyocytes are so far the best characterized cells showing coupling with the recipient myocardium even on the ultrastructural level [27]. Therefore, they can serve as a good model for the identification of factors enabling generation of pacemaker cells from early stage atrial or ventricular cardiomyocytes.

These results may then be transferred to cells other than embryonic or fetal cardiomyocytes such as embryonic and adult stem cell-derived cardiomyocytes, which after in vitro modulation could act as potential biological pacemakers following engraftment into the wall of the atrium or ventricle [24].

Acknowledgments. This study was supported by the Deutsche Forschungs gemeinschaft (Ho 2146/3-1).

References

1. Alcolea S, Jarry-Guichard T, de Bakker J, et al (2004) Replacement of connexin40 by connexin45 in the mouse: impact on cardiac electrical conduction. Circ Res 94:100–109
2. Brown HF (1982) Electrophysiology of the sinoatrial node. Physiol Rev 62:505–530
3. Cerbai E, Pino R, Porciatti F, et al (1997) Characterization of the hyperpolarization activated current, I_f, in ventricular myocytes from human failing heart. Circulation 95: 568–571
4. Coppen SR, Kaba RA, Halliday D, et al (2003) Comparison of connexin expression patterns in the developing mouse heart and human foetal heart. Mol Cell Biochem 242: 121–127
5. Delorme B, Dahl E, Jarry-Guichard T, et al (1995) Developmental regulation of connexin 40 gene expression in mouse heart correlates with the differentiation of the conduction system. Dev Dyn 204:358–371
6. DiFrancesco D (1991) The contribution of the 'pacemaker' current (if) to generation of spontaneous activity in rabbit sino-atrial node myocytes. J Physiol 434:23–40
7. DiFrancesco D (1993) Pacemaker mechanisms in cardiac tissue. Annu Rev Physiol 55:455–472
8. DiFrancesco D, Tortora P (1991) Direct activation of cardiac pacemaker channels by intracellular cyclic AMP. Nature 351:145–147
9. Er F, Larbig R, Ludwig A, et al (2003) Dominant-negative suppression of HCN channels markedly reduces the native pacemaker current I(f) and undermines spontaneous beating of neonatal cardiomyocytes. Circulation 107:485–489
10. Gassanov N, Er F, Zagidullin N, et al (2004) Endothelin induces differentiation of ANP-EGFP expressing embryonic stem cells towards a pacemaker phenotype. FASEB J 18:1710–1712
11. Gochberg SH (1964) Congenital heart block. Am J Obstet Gynecol 88:238–241
12. Hoppe UC, Beuckelmann DJ (1998) Characterization of the hyperpolarization-activated inward current (I_f) in isolated human atrial myocytes. Cardiovasc Res 38:788–801
13. Hoppe UC, Jansen E, Südkamp M, et al (1998) A hyperpolarization-activated inward current (I_f) in ventricular myocytes from normal and failing human hearts. Circulation 97:55–65
14. Klug MG, Soonpaa MH, Koh GY, et al (1996) Genetically selected cardiomyocytes from differentiating embryonic stem cells form stable intracardiac grafts. J Clin Invest 98: 216–224
15. Kusumoto FM, Goldschlager N (1996) Cardiac pacing. N Engl J Med 334:89–97
16. Meyer D, Birchmeier C (1995) Multiple essential functions of neuregulin in development. Nature 378:386–390
17. Miake J, Marban E, Nuss HB (2002) Biological pacemaker created by gene transfer. Nature 419:132–133

18. Michaelsson M, Engle MA (1972) Congenital complete heart block: an international study of the natural history. Cardiovasc Clin 4:85–101
19. Moorman AF, de Jong F, Denyn MM, et al (1998) Development of the cardiac conduction system. Circ Res 82:629– 644
20. Muller M, Fleischmann BK, Selbert S, et al (2000) Selection of ventricular-like cardiomyocytes from ES cells in vitro. FASEB J 14:2540–2548
21. Potapova I, Plotnikov A, Lu Z, et al (2004) Human mesenchymal stem cells as a gene delivery system to create cardiac pacemakers. Circ Res 94:952–959
22. Qu J, Plotnikov AN, Danilo P Jr, et al (2003) Expression and function of a biological pacemaker in canine heart. Circulation 107:1106–1109
23. Rentschler S, Zander J, Meyers K, et al (2002) Neuregulin-1 promotes formation of the murine cardiac conduction system. Proc Natl Acad Sci USA 99:10464–10469
24. Ruhparwar A, Haverich A (2003) Prospects for biological cardiac pacemaker systems. Pacing Clin Electrophysiol 26:2069–2071
25. Ruhparwar A, Tebbenjohanns J, Niehaus M, et al (2002) Transplanted fetal cardiomyocytes as cardiac pacemaker. Eur J Cardiothorac Surg 21:853–857
26. Schulze-Bahr E, Neu A, Friederich P, et al (2003) Pacemaker channel dysfunction in a patient with sinus node disease. J Clin Invest 111:1537–1545
27. Soonpaa MH, Koh GY, Klug MG, et al (1994) Formation of nascent intercalated disks between grafted fetal cardiomyocytes and host myocardium. Science 264:98–101
28. Wainger BJ, DeGennaro M, Santoro B, et al (2001) Molecular mechanism of cAMP modulation of HCN pacemaker channels. Nature 411:805–810

List of Contributors

Jacques M.T. de Bakker
Department of Experimental Cardiology,
Heart Failure Research Center,
Academic Medical Center,
Meibergdreef 9, 1105AZ
Amsterdam, The Netherlands
j.m.debakker@amc.uva.nl

Antonio Zaza
Dipartimento di Biotecnologie e Bioscienze,
Università di Milano Bicocca, Milano, Italy
antonio.zaza@unimib.it

Tobias Opthof
Experimental and Molecular Cardiology Group, Academic Medical Center,
Meibergdreef 9, 1105 AZ Amsterdam, The Netherlands
t.opthof@med.uu.nl

Eduardo Marbán
Institute of Molecular Cardiobiology,
Division of Cardiology,
Johns Hopkins University School of Medicine, 858 Ross Bldg,
720 Rutland Ave, Baltimore,
MD 21205, USA
marban@jhmi.edu

Hee Cheol Cho
Institute of Molecular Cardiobiology,
Division of Cardiology,
Johns Hopkins University School of Medicine, 858 Ross Bldg,
720 Rutland Ave, Baltimore,
MD 21205, USA

Traian M. Anghel
Department of Medicine, Section of Cardiology,
University of Illinois at Chicago,
840 South Wood Street,
M/C 715, Chicago, IL 60612, USA

Steven M. Pogwizd
Department of Medicine, Section of Cardiology,
University of Illinois at Chicago,
840 South Wood Street,
M/C 715, Chicago, IL 60612, USA
spogwizd@uic.edu

Michael R. Rosen
Department of Pharmacology,
Center for Molecular Therapeutics,
College of Physicians and Surgeons of Columbia University,
630 West 168 Street, PH7West-321,
New York, NY 10032, USA
mrr1@columbia.edu

Peter R. Brink
Departments of Physiology and Biophysics,
Institute of Molecular Cardiology,
SUNY Stony Brook,
Stony Brook, NY, USA

Ira S. Cohen
Department of Pharmacology, Center for Molecular Therapeutics,
College of Physicians and Surgeons of Columbia University,
630 West 168 Street, PH7West-321,
New York, NY 10032, USA

Richard B. Robinson
Department of Pharmacology, Center for Molecular Therapeutics,
College of Physicians and Surgeons of Columbia University,
630 West 168 Street, PH7West-321,
New York, NY 10032, USA

Gerard J.J. Boink
Department of Clinical and Experimental Cardiology,
Academic Medical Center, University of Amsterdam,
Meibergdreef 9, 1105 AZ Amsterdam,
The Netherlands

Jurgen Seppen
Liver Center, Academic Medical Center,
University of Amsterdam,
Amsterdam, The Netherlands

Hanno L. Tan
Department of Clinical and Experimental Cardiology,
Academic Medical Center, University of Amsterdam,
Meibergdreef 9, 1105 AZ Amsterdam,
The Netherlands
h.l.tan@amc.uva.nl

Teun P. de Boer
Department of Medical Physiology,
Heart Lung Center Utrecht,
University Medical Center Utrecht,
Yalelaan 50, 3584 Utrecht,
The Netherlands

Toon A.B. van Veen
Department of Medical Physiology,
Heart Lung Center Utrecht,
University Medical Center Utrecht,
Yalelaan 50, 3584 Utrecht, The Netherlands

Marien J.C. Houtman
Department of Medical Physiology,
Heart Lung Center Utrecht,
University Medical Center Utrecht,
Yalelaan 50, 3584 Utrecht,
The Netherlands

John A. Jansen
Department of Medical Physiology,
Heart Lung Center Utrecht,
University Medical Center Utrecht,
Yalelaan 50, 3584 Utrecht,
The Netherlands

Shirley C.M. van Amersfoorth
Experimental and Molecular Cardiology Group, Academic Medical Center,
Amsterdam, The Netherlands

Pieter A. Doevendans
Department of Cardiology, University Medical Center Utrecht, Utrecht,
The Netherlands

Marc A. Vos
Department of Medical Physiology,
Heart Lung Center Utrecht,
University Medical Center Utrecht,
Yalelaan 50, 3584 Utrecht, The Netherlands

Marcel A.G. van der Heyden
Interuniversity Cardiology Institute of the Netherlands, Utrecht,
The Netherlands
m.a.g.vanderheyden@umcutrecht.nl

Ronald W. Joyner
Department of Pediatrics, Emory University
School of Medicine, 2015 Uppergate Drive NE,
Atlanta, GA 30322, USA
rjoyner@cellbio.emory.edu

Ronald Wilders
Department of Physiology, Academic Medical Center,
University of Amsterdam, Meibergdreef 15,
1105 AZ Amsterdam, The Netherlands
r.wilders@amc.uva.nl

Mary B. Wagner
Department of Pediatrics, Emory University
School of Medicine, 2015 Uppergate Drive NE,
Atlanta, GA 30322, USA

Yuichi Tomita
Department of Regenerative Medicine and Advanced Cardiac Therapeutics,
Keio University School of Medicine,
Institute of Integrated Medical Research,
35 Shinanomachi, Shinjuku-ku, Tokyo 160-8582, Japan

Shinji Makino
Department of Regenerative Medicine and Advanced Cardiac Therapeutics,
Keio University School of Medicine,
Institute of Integrated Medical Research,
35 Shinanomachi, Shinjuku-ku, Tokyo 160-8582, Japan

Daihiko Hakuno
Department of Regenerative Medicine and Advanced Cardiac Therapeutics,
Keio University School of Medicine,
Institute of Integrated Medical Research,
35 Shinanomachi, Shinjuku-ku, Tokyo 160-8582, Japan

Naoichiro Hattan
Department of Regenerative Medicine and Advanced Cardiac Therapeutics,
Keio University School of Medicine,
Institute of Integrated Medical Research,
35 Shinanomachi, Shinjuku-ku, Tokyo 160-8582, Japan

Kensuke Kimura
Department of Regenerative Medicine and Advanced Cardiac Therapeutics,
Keio University School of Medicine,
Institute of Integrated Medical Research,
35 Shinanomachi, Shinjuku-ku, Tokyo 160-8582, Japan

Shunichiro Miyoshi
Cardiology Division, Department of Internal Medicine,
Keio University School of Medicine,
35 Shinanomachi, Shinjuku, Tokyo 160-8582, Japan

Mitsushige Murata
Department of Regenerative Medicine and Advanced Cardiac Therapeutics,
Keio University School of Medicine,
Institute of Integrated Medical Research,
35 Shinanomachi, Shinjuku-ku, Tokyo 160-8582, Japan

Masaki Ieda
Department of Regenerative Medicine and Advanced Cardiac Therapeutics, Keio University School of Medicine, Institute of Integrated Medical Research,
35 Shinanomachi, Shinjuku-ku, Tokyo 160-8582, Japan

Keiichi Fukuda
Department of Regenerative Medicine and Advanced Cardiac Therapeutics, Keio University School of Medicine, Institute of Integrated Medical Research,
35 Shinanomachi, Shinjuku-ku, Tokyo 160-8582, Japan
kfukuda@sc.itc.keio.ac.jp

Arjang Ruhparwar
Department of Cardiac Surgery, University of Heidelberg, 69120 Heidelberg, Germany
Arjang.Ruhparwar@med.uni-heidelberg.de

Fikret Er
Department of Internal Medicine III, University of Cologne, Kerpener Str. 62, 50937
Cologne, Germany

Ulrich Martin
Department of Thoracic and Cardiovascular Surgery, Hannover Medical School,
Carl-Neuberg-Str. 1, 30625 Hannover, Germany

Kristin Radke
Department of Thoracic and Cardiovascular Surgery, Hannover Medical School,
Carl-Neuberg-Str. 1, 30625 Hannover, Germany

Ina Gruh
Department of Thoracic and Cardiovascular Surgery, Hannover Medical School,
Carl-Neuberg-Str. 1, 30625 Hannover, Germany

Michael Niehaus
Department of Cardiology and Angiology, Hannover Medical School, Carl-Neuberg-Str 1, 30625 Hannover, Germany

Matthias Karck
Department of Cardiac Surgery, University of Heidelberg, 69120 Heidelberg, Germany

Axel Haverich
Department of Thoracic and Cardiovascular Surgery, Hannover Medical School,
Carl-Neuberg-Str. 1, 30625 Hannover, Germany

Uta C. Hoppe
Department of Internal Medicine III, University of Cologne, Kerpener Str. 62, 50937
Cologne, Germany

MIX
Papier aus verantwortungsvollen Quellen
Paper from responsible sources
FSC® C105338

If you have any concerns about our products,
you can contact us on
ProductSafety@springernature.com

In case Publisher is established outside the EU,
the EU authorized representative is:
Springer Nature Customer Service Center GmbH
Europaplatz 3, 69115 Heidelberg, Germany

Printed by Libri Plureos GmbH
in Hamburg, Germany